SCIENCE IN BUILDING

for Craft Student and Technician

3

MATERIALS

E. C. ADAMS

*Lecturer in Building and Craft Sciences at
Crawley College of Further Education*

HUTCHINSON EDUCATIONAL

HUTCHINSON EDUCATIONAL LTD
178–202 Great Portland Street, London W1

London Melbourne Sydney Auckland
Wellington Johannesburg Cape Town
and agencies throughout the world

First published 1969
Second revised edition June 1971

*This book has been set in Times type, printed in Great Britain
on smooth wove paper by Anchor Press, and
bound by Wm. Brendon, both of Tiptree, Essex*

ISBN 0 09 108140 8 (cased)
0 09 108141 6 (paper)

CONTENTS

LIST OF EXPERIMENTS

USEFUL TABLES AND DATA

PREFACE TO FIRST EDITION

This third and final part of *Science in Building for Craft Student and Technician* aims to present a self-contained and concise treatment of the science and technology of the materials of construction for senior craft students and technicians at all levels, as well as those following a specialist materials course.

In view of the imminence of metrication throughout British industry, emphasis has been given in this volume to the introduction of metric or S I (International) units wherever possible. Where product dimensions are necessarily in Imperial units, the metric equivalents have been included. Worked examples are either wholly in metric, or S I, terms, or, to allow for comparison and familiarisation with both systems, in Imperial and metric (or S I) units. For example, *dimensions* may be in both inches or millimetres, units of *mass* in pounds and in kilogrammes, *force* in Imperial and metric gravitational units (e.g. ton f and kgf) as well as S I absolute units (N or kN), and *stress* or *pressure* in the corresponding units of force per unit area. In addition, a Table of conversion factors is given on page 325.

The use of the *hectobar* as a unit of stress is current in some Continental practice, but it is uncertain how far this will influence future trends in Great Britain. Its conversion factors are included in the Table on page 325.

The precise changes to be made in the current British Standard test sieve sizes (BS 410:1962)* under the programme of metrication were, at the time this book went to press, uncertain, but the revision of this British Standard will take place in due course. A list of the

---

* Metric edition issued 1969.

currently proposed *preferred* (I.S.O. R565*) and *supplementary* sizes to be used in the new edition of BS 410 are given in Table 6(b) of this book. In conjunction with Table 6(a) this shows that for each of the common *inch* and *BS numbered* sieve sizes of BS 410:1962 used in the Building Industry, there is one of the exact same, or very similar, aperture dimension. Accordingly, and because relevant existing specifications must remain in use for some time, in this book the current inch and BS numbered sieve aperture sizes are given, and their metric equivalent stated. Any future changes in the standard sieve aperture sizes will not affect the principles illustrated by experiments included in the text, nor the methods of the worked examples given.

In the section on Concrete (Chapter 8) I have chosen to deal with *nominal* mixes in a traditional way, as well as including *standard* and *designed* mixes. This I considered appropriate not only because the use of standard mixes (*specified* by weight) as an alternative to nominal mixes (*specified* by volume but often batched by weight) is a comparatively recent innovation, but also because it is likely that nominal mixes will remain in use to some extent, especially for mixes with lightweight aggregates. Acknowledgement is made to the following as sources of information and for permission to use published data:

The British Standards Institution (British Standards House, 2 Park Street, London, W.1)—see also 'Note on the British Standards Institution on page 13.

The Cement and Concrete Association (52 Grosvenor Gardens, London, S.W.1) for data in Fig 28 'Table of aggregate-cement ratios for four degrees of concrete workability, with different values of water-cement ratio'.

Pilkington Brothers Limited (St. Helens, Lancashire) for technical comment on Chapter 17 'Glass'.

Timber Research and Development Association for permission to use data published in 'Timber Information' series Ref. No. 25 'Key to the microscopical identification of the principal softwoods used in Great Britain' as a basis for Table 33 on page 320.

*International Organisation for Standardisation'. I.S.O. Recommendation R565 'Woven wire cloth and perforated plates in test sieves—nominal sizes of apertures' (1st Edition, April 1967).

E.C.A.

NOTE ON THE
BRITISH STANDARDS INSTITUTION

Throughout this book there is frequent reference to the relevant British Standards and British Standard Codes of Practice published by the British Standards Institution (see list on page 327).

The British Standards form a convenient basis for the specification of materials and products, and of the testing procedures for quality control.

Codes of Practice give general information, guidance and recommendations on the use of materials and methods of construction.

These publications are subject to frequent amendments and periodic revision in order to keep pace with technological developments, and the reader is advised to consult the current issues whenever full, up-to-date information is needed. Publication dates of British Standards and British Standard Codes of Practice referred to in this book are given on page 327.

The following extracts are reproduced by permission of the British Standards Institution, 2 Park Street, London, W.1.

Table 5, page 305 (Minimum weight of sample for sieve analysis), and Table 7, page 306 (Maximum weight to be retained at the completion of sieving) from *BS 812:1967* '*Methods for sampling and testing of mineral aggregates, sands and fillers*'.

Table 34, page 323 (Grading of plywood manufactured from tropical hardwoods), and Table 35, page 324 (Bonding of plywood made from tropical hardwoods) from *BS 1455: 1963*, '*Specification for plywood manufactured from tropical hardwoods*'.

Table 36, page 324 (Classification of glass for glazing), from

BS 952:1964 'Classification of glass for glazing and terminology for work on glass'.

Experiments

A number of laboratory experiments in this book follow closely the methods used in British Standard tests and it is important to remember that examination syllabuses frequently require outline knowledge of these. Their only purpose is to illustrate the methods used and the properties of the materials concerned, and results obtained are intended to be confined to this purpose.

Tests intended to determine the compliance or otherwise of materials or products with the requirements of the British Standard must be carried out strictly as specified in the current British Standard concerned.

1

CEMENTS

PORTLAND CEMENTS

By far the commonest type of cement used in the building industry in Great Britain is the cement known as *ordinary Portland*, which got its name originally because of its resemblance, when set, to the natural Portland limestone from Dorset. The principal raw materials for making this cement are chalk (or limestone) and clay, both of which fortunately occur in extensive natural deposits in Great Britain.

The essential property of Portland cement is that it hardens by a chemical action when mixed with water. This process is known as *hydration*. A paste of cement and water first undergoes gradual stiffening, called *setting*, but the hydration continues long after the paste has become rigid, resulting in a progressive increase in strength, called *hardening*.

In addition to ordinary Portland cement there are other types. Some are made from the same raw materials but manufactured to give slightly different properties, such as different rates of setting or hardening, or greater resistance to chemical attack. These modified forms of cement, and ordinary Portland, together make up the Portland *group* of cements. There are also a number of *special* cements, including blends of Portland with other cementitious materials, as well as cements made from different raw materials. We shall first consider the Portland cements.

Manufacture

There are two methods of manufacture in use for Portland cements, the *wet* process and the *dry* process. The wet process was developed for the softer clays and the dry process was developed for the harder clays and shales.

For the wet process the materials are chalk and clay, the proportions being roughly three parts of chalk (or limestone) to one part of clay; water is added and the substance is then reduced to a slurry, first by crushing then by grinding in wash mills. The slurry then passes to storage tanks, where it is kept agitated to prevent settlement. From there the slurry is pumped to the inlet end of a rotary kiln, which is a large steel cylinder, lined with refractory blocks. The kiln revolves slowly on an axis slightly inclined to the horizontal, causing the contents to gravitate towards the lower end. The complete process takes several hours. Pulverised coal (or oil fuel) is blown in at the lower end of the kiln and burns continuously, causing the materials in the slurry to combine chemically. For this to occur a temperature of about 1250°C must be reached at the hottest part of the kiln (near its lower end). During the kilning the slurry first loses its free water content, then becomes dehydrated. The calcium carbonate decomposes to calcium oxide (quicklime), and finally there is chemical combination between the quicklime and the compounds in the clay to form the cement. The cement leaves the kiln in the form of small nodules, *cement clinker*. The clinker is cooled by being passed through a ventilated rotating cylinder, similar to the kiln itself but of shorter length, and is finally ground in a tube mill to form the familiar grey powder, ordinary Portland cement. During the final grinding process a small quantity of gypsum (2–5%) is added to prevent the cement from setting too quickly after water has been added.

In the dry process the materials are crushed, dried, then ground in a tube mill to form a powdered mixture which is then burnt from the dry state. This method uses less fuel than the wet process.

Composition

Ordinary Portland cements are of variable composition, but a typical chemical analysis would show roughly 60% of lime (CaO), 20% of silica (SiO_2), 10% alumina (Al_2O_3), 3% oxides of iron and small amounts of other materials. These materials are present mainly in chemical combination, the most important compounds being the silicates and aluminates of calcium. This is further illustrated in Fig 1.

ORDINARY PORTLAND CEMENT			
A–RAW MATERIALS			

ARGILLACEOUS (Clay, shales)	I PART	3 PARTS	CALCAREOUS (Chalk, limestone)

or
MARL

Note: only gypsum and water may be added after burning

B–CHIEF CHEMICAL CONSTITUENTS IN RAW MATERIALS			
LIME 65%	SILICA 20%	ALUMINA 5%	IRON OXIDE 3%
High lime content increases setting time but gives high early strength. Too little reduces strength. Excess if hard burnt causes unsoundness	High silica content increases setting time and strength	High alumina content reduces setting time and increases strength	Imparts grey colour

C–CHIEF COMPOUNDS IN FINISHED PRODUCT (POWDER)			
CALCIUM SILICATES 75%		CALCIUM ALUMINATES 25%	
Tricalcium silicate C_3S Early strength High heat More in R H Less in L H	Dicalcium silicate C_2S Later strength Little heat	Tricalcium aluminate C_3A Little strength Great heat Unstable, decomposing to lime in the presence of air or water. Attacked by sulphates to form calcium sulphoaluminate. Less in S R and L H	Tetra-calcium alumino ferrite C_4AF Little strength Little heat

D–COMPOUNDS IN SET CEMENT
Their precise nature has not yet been confirmed. They are believed to consist of interlacing needles of the hydrated crystals of silicates and aluminates, together with hydrated forms of the iron compounds and free hydrated lime

Note: all proportions are variable and typical values only are given here
R H – Rapid Hardening Portland cement
L H – Low Heat Portland cement
S R – Sulphate Resisting Portland cement

Fig I The chemistry of ordinary Portland cement

Hydration

When Portland cement is mixed with water the two combine in the chemical process known as *hydration*, and a crystalline structure develops. It is the growth of the crystals of the various compounds which causes the *setting* and *hardening* of the cement paste. Hydration is accompanied by the release of small amounts of uncombined calcium hydroxide—$Ca(OH)_2$—known as the *free lime* content of the hardened cement. Provided that the necessary water is available, hydration, and therefore strength gain, continues indefinitely, although at a progressively reducing rate. This explains the need to keep concrete damp for a time after placing, a procedure known as *curing*. During hydration, heat is given off.

Rapid-hardening Portland cement

If Portland cement is ground finer than normal in the final manufacturing stage, the rate at which hydration will occur is speeded up, and the cement paste hardens more rapidly. This finer grinding is the principal feature of *rapid-hardening* Portland cement, which is the same as ordinary Portland, except for a very minor adjustment of its composition made after chemical analysis at the slurry stage if found necessary.

TESTS FOR ORDINARY AND RAPID-HARDENING CEMENTS

Full requirements and tests for both these cements are given in British Standard 12 (usually written as 'BS 12') 'Portland cement (ordinary and rapid-hardening)'.

The tests are for fineness and chemical composition of the cement powder, strength and soundness of hardened specimens, and setting time of cement paste. A summary of the testing methods follows, and current BS requirements are given in Table 1, page 301.

1 Fineness

Fineness is important because of its effect on the rate of hydration, and the need for a smooth texture. The test was originally carried out to determine the fraction of material passing a fine-mesh sieve (BS 170/90 microns) but this was dropped in favour of a test which assessed the overall fineness by measuring the surface area of the particles per gram of cement (cm²/g), known as its *specific surface*. Details of the special *permeability apparatus*, which measures the rate of flow of air through a bed of compacted cement powder, under standardised conditions, are given in BS 12.

For routine control tests, other forms of apparatus, known as *flowmeters*, are available which are simpler to use and can be calibrated against the BS type.

2 *Chemical composition*

This test is to ensure that the cement conforms to a reasonable standard of chemical composition and is free of excessive harmful impurities. The test is made by chemical analysis, and deals with lime content, the permissible amounts of magnesia and sulphuric anhydride, insoluble residue and loss on ignition.

3 *Strength*

Hardened specimens are tested after specified curing. In the *mortar cube test*, cubes of 50 cm² face are prepared by weighing up 1 part of cement to 3 parts of a standard sand, and mixing them with a fixed proportion of water (water:cement ratio 0·40). The cubes are vibrated in a special machine to remove entrapped air and are then tested for crushing strength after specified curing.

The alternative *concrete cube test* uses cubes size 4 in (or 100 mm) made from a mix of 1 part of cement to approximately 6 parts of approved aggregate (sand and gravel), varying according to the aggregate specific gravities, with a fixed amount of water (water: cement ratio 0·60).

A *tensile strength* test made on *mortar briquettes* is at present an alternative to the crushing strength tests for rapid-hardening Portland cement only. The briquettes are cast in figure-eight moulds (Fig 2) to allow a grip by the jaws of the tensile testing machine. Mix proportions are, by weight, 1 part of cement, to 3 parts of a standard sand with a fixed amount of water (water:cement ratio 0·32).

4 *Setting time*

The Vicat apparatus (Fig 3) is used for this test. The time for a neat cement paste of standard consistency to reach two defined conditions

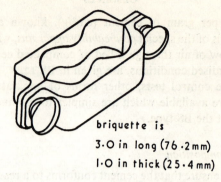

briquette is

3·0 in long (76·2mm)

1·0 in thick (25·4mm)

Fig 2 A cement tensile briquette mould

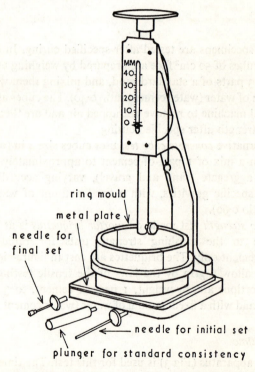

MM
40
30
20
10
0

ring mould

metal plate

needle for
final set

needle for initial set

plunger for standard consistency

Fig 3 Vicat setting test apparatus

of stiffness is determined by periodically checking the penetration of a needle into it. The Vicat apparatus has three needle attachments:

[a] a cylindrical plunger, 10 mm diameter with flat end, for use in establishing the amount of water to produce a cement paste of standard consistency

[b] a square, or round, needle of cross-sectional area 1 mm², used to determine the *initial* setting time

[c] a needle of the same section as that used for the initial setting determination, but having an attached collar with its lower rim 0·5 mm from the lower end of the needle—used to determine the *final* setting time

5 *Soundness*

Unsoundness in a material denotes the presence of impurities liable to react chemically with moisture with resulting expansion, which can cause cracking, spalling or disintegration. The simplest test for unsoundness is to steam, or boil, a specimen to accelerate the action. For cements the *Le Chatelier* apparatus is used (Fig 4),

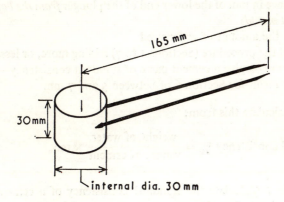

Fig 4 Le Chatelier mould for soundness tests

which is a split-ring mould with attached pointers. A hardened specimen of neat cement paste, cast in the mould, is boiled for 1 hour in water and any expansion is noted by measuring relative movement of the pointers.

EXPERIMENT 1   Consistency of a standard cement paste.

Apparatus Vicat apparatus to BS 12, with plunger attachment and ring mould (Fig 3)—gauging trowels—measuring cylinder (250 ml)—stopclock—mixing board (e.g. glass or metal plate)—physical balance (metric)

Specimen Portland cement (ordinary or rapid-hardening)

Method

1 Weigh out 400 g of cement powder and place it on the mixing board

2 Measure out exactly 200 ml of water

3 Start the stopclock and mix water with the cement to produce a fairly stiff paste (a little more than half the water will be needed). Mix for 4 minutes

4 Fill the paste into the mould, striking it level with the top edge and smoothing the surface with a trowel

5 Locate the mould beneath the plunger, lower the plunger to contact the surface of the paste, then release it

6 Observe the depth of penetration of the plunger (the indicator gives the distance in mm of the lower end of the plunger *from the bottom of the Vicat mould*)

7 Check the amount of water used

8 Repeat the procedure (Sections 1 to 7) adding more, or less, water as necessary to give a cement paste of standard consistency—that is to give a result for Section 6 of between 5 and 7 mm

Result Calculate this from:

$$\text{Standard consistency } \% = \frac{\text{weight of water}}{\text{weight of cement}} \times 100$$

Example 1 Calculate the standard consistency of a cement from the following test results:

Weight of cement used . . . 400 g

Water required for plunger to penetrate to 5–7 mm from baseplate 120 ml

$$\text{Standard consistency} = \frac{120}{400} \times 100 = 30\%$$

EXPERIMENT 2 Setting times of Portland cement

Note The time for initial set may be up to several hours. If the full time is not available the experiment can be modified to the form of a demonstration only

Preliminary test The standard consistency of the cement used must be known (see Experiment 1)

Apparatus The same apparatus which was used in Experiment 1, with the addition of the two remaining Vicat needles

Specimen Portland cement (ordinary or rapid-hardening)

Method
[*a*] Initial setting time
1 Fit the needle without the collar in the Vicat apparatus
2 Prepare a cement paste of standard consistency, using 400 g of cement, and water as determined by a preliminary test. Do not forget to start the stopclock as soon as the water is added, and mix for 4 minutes
3 Fill the paste into the Vicat mould and strike it level
4 Position the mould beneath the Vicat needle, lower the needle to contact the surface of the paste, then release it, observing the degree of penetration
5 Repeat the procedure of Section 4 at intervals, with the needle at different points of the surface, until penetration is not beyond a point 5 mm from the baseplate. The time to reach this condition from the commencement of mixing gives the initial setting time (in hours and minutes)

[*b*] Final setting time
Immediately following the initial setting and with the stopclock continuing:
1 Change the needle in the Vicat apparatus for that with the collar attachment
2 Continue as in Section 4, until the collar first fails to make an impression while the tip of the needle still does. The time it has taken to reach this condition from the time when the water was added gives the final setting time.

Example 2 The standard consistency of a sample of Portland cement is 30%. Calculate the amount of water to be mixed with 50 g of the cement to give a paste of standard consistency.

$$\text{Water required} = \frac{30}{100} \times 50$$

$$= 15 \text{ g (or 15 ml)}$$

EXPERIMENT 3 Soundness of a Portland cement

Preliminary test The standard consistency of the cement used must be known (see Experiment 1)

Apparatus Le Chatelier moulds to BS 12 (Fig 4)—gauging trowel and mixing board—plates of thin sheet glass, 50 mm square and small lead or brass weights (about 100 g)—burette (or 25 ml measuring cylinder)—dish to take moulds and boiling can or beaker—vaseline

Method (for one specimen)

1 Check that the split of the mould is not more than 0·5 mm wide (adjust if necessary), and apply vaseline thinly to the mould and glass-sheet surfaces

2 Take 50 g of cement and mix it with water to give a paste of stand-ard consistency (see Example 2)

3 Place the mould on a glass plate and fill it with cement paste (closing the split temporarily by gentle pressure between two fingers)

4 Cover the mould by another glass plate, put a small weight on it and immerse the mould and specimen in water for 24 hours

5 After immersion, measure the distance, in mm, between the ends of the pointers

6 Submerge the specimen (with mould) in water, heat to reach boiling point in 25 to 30 minutes, and boil for 1 hour

7 Allow the specimen to cool, then re-measure the distance between the pointer ends

Result This is the difference between the two measurements (Sections 5 and 7). Table 1, page 301, gives the BS test limits.

EXPERIMENT 4 The specific gravity of a cement powder

Apparatus Specific gravity bottle—laboratory chemical balance—vacuum vessel (e.g. desiccator and pump)—thin paraffin oil (e.g. re-distilled kerosine, BP 200–240°C)—drying cloth

Specimen Cement powder

Method

1 Weigh the empty bottle, dry and with its stopper

2 Add cement powder until the bottle is about one-third full, then re-weigh it

3 Add oil to half fill the bottle, then place it in the vacuum vessel for half an hour to remove the air

4 Add oil until the bottle is full, insert the stopper and wipe off the excess liquid

5 Wipe the outside of the bottle dry and weigh it

6 Empty the bottle and rinse it with oil; now fill the bottle with oil, wipe it dry, and weigh it

7 Empty the bottle, thoroughly cleanse it with warm water containing detergent, then rinse it with cold water and fill it with distilled water. Now wipe dry the outside, and weigh it again. This is not necessary for a bottle already calibrated

Result The calculation for this is shown by the worked example which follows.

Example 3 It is required to calculate the specific gravity of a cement from test data obtained by the method of Experiment 4.

The calculation is set out below, with data from the experiment and the Section number noted.

Mass of bottle+cement (2)	= 60·89 g
Mass of bottle only (1)	= 39·50 g
Mass of cement used (2)−(1)	= 21·39 g
Mass of bottle+cement+oil to fill (4)	=134·26 g
Mass of oil used (4)−(2)	= 73·37 g (A)
Mass of bottle+oil to fill (6)	=118·30 g
Mass of bottle only (1)	= 39·50 g
Mass of oil only to fill bottle (6)−(1)	= 78·80 g (B)
Mass of bottle+water to fill (7)	=139·51 g
Mass of bottle only (1)	= 39·50 g
Mass of water only to fill bottle (7)−(1)	=100·01 g (C)
Specific gravity of oil=B÷C	= 0·788 (D)
Mass of oil of same volume as cement	=B−A
	= 5·43 g

Mass of water of same volume as cement $= \dfrac{B-A}{D}$

$$= 5 \cdot 43 \div 0 \cdot 788$$
$$= 6 \cdot 89 \text{ g}$$

s.g. of cement $= \dfrac{\text{mass of cement}}{\text{mass of an equal volume of water}}$

$$= 21 \cdot 39 \div 6 \cdot 89 \qquad\qquad = 3 \cdot 10$$

MODIFIED PORTLAND CEMENTS

A number of different types of Portland cement with modified properties to suit particular circumstances are available. While these may have different rates of strength gain, it is useful to remember that they will all in time eventually reach about the same strength.

Rapid-hardening cement

You have already seen that the main feature of the cement is its finer grinding than ordinary Portland, and its main use is in concrete work where greater early strengths are required than would be obtained with ordinary Portland.

Extra rapid-hardening cement

The rate of hardening of this cement is increased even further by adding an accelerator to rapid-hardening cement. The usual method is to add 2% of calcium chloride to the cement during its final grinding in manufacture. Alternatively, the same effect is obtained by adding the calcium chloride to rapid-hardening cement at the time of mixing, preferably in solution and added to the mixing water, for good dispersion in the mix. This cement is used mainly for concreting work in winter conditions to counteract the reduced rate of hydration at low temperatures. The cement is also quick-setting, so the concrete should normally be placed in its final position within 20 minutes of leaving the mixer.

An increase in the rate of setting and hardening of cement is accompanied by an increase in the rate at which the heat of hydration is given off. In cold weather this lessens the likelihood of frost damage to new work. On the other hand, in mass concrete structures like dams, the heat of hydration is not easily lost and can cause over-heating internally, resulting in cracking and even disintegration. For such cases, a low-heat Portland cement can be used.

Low-heat Portland cement

This cement is of modified composition to give a low rate of heat evolution. BS 1370:1958 limits the heat of hydration of this cement to not more than 60 calories per gram at 7 days, nor more than 70 calories per gram at 28 days. A typical ordinary Portland cement evolves about 100 calories per gram in 28 days.

Sulphate resisting cement

This is a modified Portland cement with improved resistance to chemical attack by *sulphates* which are salts found in sea-water, certain ground waters, and (as impurities) in some building materials. Sulphates in solution can cause softening, and considerable expansion of cement-based materials.

White cement

This is made by using *china clay* (kaolin), which is a pure clay, free of iron oxides. Iron oxides give ordinary Portland cement its grey colour.

Coloured cements

These are made by blending inert pigment powders with white, or grey, cement, depending on the colour required. The pigments may be added at the works, or, alternatively, by the user.

Water-repellent cement

This is produced by intergrinding the Portland cement clinker with a small proportion of a water-repellent, such as gypsum with tannic acid, or certain metallic soaps. These cements are used in water-repellent renderings and in base-coats where the background is of uneven suction (e.g. no-fines concrete), especially where a coloured finish coat is to be applied, in order to avoid patchiness, as well as in dense concrete.

Hydrophobic cements

These are treated in manufacture so that the cement powder does not readily absorb moisture, and can be stored for considerable periods in damp or humid conditions without deteriorating.

CEMENTS OTHER THAN PORTLAND

These include Portland blast-furnace cement, masonry cement, super-sulphated cement and high-alumina cement.

Portland blast-furnace cement to BS 146 is an interground mixture of up to 65% of basic blast-furnace slag and ordinary Portland cement. It is a *low-heat* cement, slower in its early hardening than ordinary Portland, but can in the long term be of equal or greater strength. Its high resistance to sulphates makes it popular for marine construction.

Masonry cement

This is produced specially to give the high workability desirable in mortars for masonry, brickwork and rendering without the addition of lime or other plasticiser by the user. It consists usually of a blend of ordinary Portland cement and finely powdered chalk or silica, possibly with the addition of a plasticising chemical.

Supersulphated cement

This is an interground mixture of 80 to 85% of basic blast-furnace slag and 10 to 15% of gypsum, with about 5% of Portland cement clinker as an activator. It is a *low-heat* cement and has a high resistance to sulphates, peaty soils and oils.

High-alumina cement (Aluminous cement)

This cement is made by heating a mixture of *bauxite* (aluminium ore) and limestone. The mixture melts at about 1600°C and is cast into 'pigs', then ground to a powder.

This cement (to BS 915) has a high resistance to most chemicals (particularly sulphates), except alkalis. It very rapidly gains strength and reaches about 80% of its full strength in only 24 hours. Continuous wet curing is essential during this period owing to the great heat generated. It must always be remembered that concrete which is made with this cement will permanently lose up to 50% or more of its maximum strength if subjected, when moist, to temperatures above about 30°C at any time during its life due to a change occurring in the chemical nature of the bond, called *reversion*. This must be taken into account structurally. High-alumina cement should only be placed in thin sections to prevent cracking from thermal expansion where the heat of hydration cannot easily disperse. The cement is widely used in refractory concrete, since it remains stable well above 1000°C, whereas Portland cement concretes are unlikely to withstand prolonged heating much above 500°C.

Summary

Portland cements are made by calcining a mixture of 3 parts of chalk or limestone (calcium carbonate) to 1 part of clay (hydrated aluminium silicates) approximately, then grinding the resulting clinker, with a small addition of gypsum to control the setting time.

Adding water to the cement causes hydration, resulting in progressive stiffening (setting), followed by hardening (strength increase), with the evolution of heat.

Rapid-hardening cement is more finely ground than ordinary Portland and may have a slightly different chemical composition. Adding an accelerator to it (2% of calcium chloride) gives an extra rapid-hardening cement. The increased rate of hydration results in quicker setting and hardening (extra rapid-hardening cement is classified as a quick-setting cement) with an increased rate of heat evolution.

Modified forms of Portland cement include low-heat, sulphate-resisting, water-repellent, hydrophobic, and white and coloured cements. Other cements include Portland blast furnace (up to 65% slag interground with ordinary Portland), masonry cement (a blend of ordinary Portland and finely ground chalk or silica, sometimes with a plasticiser), supersulphated cement (a blend of 80–85% B.F. slag, 10–15% gypsum and Portland cement) and high-alumina cement (from bauxite and limestone).

BS 12 tests for ordinary and rapid-hardening Portland cements are for fineness, chemical composition, strength, setting time and soundness.

2

GYPSUM BUILDING PLASTERS

Gypsum building plasters are also known as *calcium sulphate* plasters, since this is the chemical name for *gypsum*, the natural rock from which they are made. It is the *dihydrate* form ($CaSO_4 . 2H_2O$). There also exist natural deposits of the anhydrous form ($CaSO_4$) called *anhydrite*, but this is little used in the plaster industry of Great Britain. Yet another name given to these plasters is *hardwall plasters*.

MANUFACTURE OF PLASTERS

Gypsum plasters are made by heating gypsum to drive off part or all of the water of crystallisation. With strong heating, all the water is driven off and anhydrous gypsum plaster ($CaSO_4$) is formed. With less heat only three-quarters of the water is removed, giving the *hemihydrate* ($2CaSO_4 . H_2O$) which, for convenience, is usually halved and written as $CaSO_4 . \frac{1}{2} H_2O$, since this clarifies the relationship between the three forms:

anhydrous	*hemihydrate*	*dihydrate*
$CaSO_4$	$CaSO_4 . \frac{1}{2} H_2O$	$CaSO_4 . 2 H_2O$

If water is added to the powdered anhydrous or hemihydrate forms, conversion to the dihydrate (by hydration) takes place and a setting action occurs. It is this action on which the plasterer depends.

Whereas pure gypsum is white, the presence of traces of impurities in the raw material leads to gypsum plasters of different colours, such as grey, pink or brown. Their use is mainly for internal finishes to walls and ceilings, either neat or with fine aggregate or lime added. The sand or lime is usually mixed in as required by the plasterer, but premixed plasters incorporating the lightweight aggregates *vermiculite* and *perlite* require only the addition of water before use.

Anhydrous plaster

The crushed rock, calcined in a kiln at 400°C or above to drive off all the water of crystallisation, is ground to a powder in a tube mill. A small amount of an accelerator (e.g. aluminium or potassium sulphate) is added, normally in solution, before calcining the powdered gypsum, to speed the setting action, which would otherwise be too slow.

Anhydrite

This plaster is prepared simply by grinding the natural anhydrite together with an accelerator.

Hemihydrate plaster

The material first obtained in the heating of gypsum is *plaster of Paris*. This is the true hemihydrate ($CaSO_4 . \frac{1}{2} H_2O$) and sets far too quickly for general use in plastering work, although it is useful for small areas and in making precast mouldings, such as cornices. To extend the setting time a *retarder* (e.g. keratin derivatives, or lime with an activator) is added during manufacture. The finished product is known as *retarded hemihydrate* plaster.

In manufacture the crushed rock is finely ground then heated for some hours at 150°C in a kettle (kiln) or open pan, or alternatively in a vessel called an *autoclave* heated by steam pipes.

SETTING OF PLASTER

The setting action of gypsum plasters results from the growth of crystals during the hydration. A slight expansion normally takes place as the setting advances and continues in the set material as the hydration proceeds and goes to completion. The setting action (hardening) of most commercial plasters is normally complete within several hours, but the hydration usually occupies a longer period and for some products may still be incomplete after several days. In general, anhydrous types have a slower, more steady, rate of hydration than hemihydrate types.

The absence of drying shrinkage means that there are no shrinkage cracks, and, in addition, a further plaster coat can follow as soon as the previous one has set without the need for a drying period.

If gypsum plasters are dried too rapidly after application the hydration may be incomplete. The residual hydration can then take place later, due to atmospheric moisture or direct wetting, and the resulting expansion can cause bond failure and flaking of plasterwork. This *delayed expansion* is less likely to occur with hemihydrate plasters than with anhydrous types (which take longer to complete their hydration) and with the latter the prevention of dryout is particularly important in the first 48 hours after application.

Although hemihydrate plasters usually take up less water of crystallisation in hydrating than anhydrous types, they need the same amount of water to produce a workable mix. The evaporation of the 'excess' water usually results in the set hemihydrate being more porous, and less hard, than the anhydrous types. Anhydrous types may exhibit a *'double set'* (an early stiffening followed by a more gradual hardening) which is perhaps due to the presence of particles not fully dehydrated in manufacture. Such plaster, if not applied before the initial stiffening, may be *retempered* (that is, more water added) only once, immediately after this occurs, and this is said to produce a *fattier* (more workable) mix. Hemihydrate plasters should not be retempered.

Plasters, like cements, must be kept dry in storage to prevent premature hydration.

EXPERIMENT 5 Preparation and setting of gypsum plaster

Apparatus Fireclay crucible with iron stand—bunsen burner—metal stirrer—mixing board—palette knife—beaker—stopclock—physical balance (metric)

Specimen Powdered gypsum (calcium sulphate dihydrate)

Materials Prepared gelatin or size solution (add 5 g of gelatin or size to 1 litre of hot water, stir to dissolve, then cool)

Method
[a] *Preliminary test*
Add water to a little of the powdered gypsum (dihydrate) and confirm that there is no setting action

[b] *Calcining gypsum*

1 Weigh out 50 g of powdered gypsum into the crucible and heat gently over a bunsen flame, stirring continuously. Steam will be given off and the plaster will appear to 'boil'. Continue heating gently until the 'boiling' ceases

2 Allow the calcined powder to cool, then divide it into two equal portions for the remaining tests

[c] *Setting of calcined plaster*

1 Place one part of the calcined gypsum on a *clean* mixing board

2 Fill the beaker with water, start the stopclock and mix water with the calcined gypsum to give a plastering consistency

3 Observe the time for the plaster to become hard (the approximate setting time)

[d] *Effect of gelatin or size solution on setting*

Using a *clean* mixing board and with the second part of the calcined gypsum, repeat the procedure of Section [c] but using the prepared solution, instead of plain water

Results Compare the setting times and draw conclusions

Note Clean mixing utensils are essential, because set plaster can act as an accelerator.

TYPES AND USES OF PLASTER

Different grades of plaster are available for different purposes; for example, *undercoat* or *browning*, *finishing* and *dual* purpose. The latter is suitable for both undercoat or final coat. Some undercoats may be of the *fibred* variety (formerly with oxhair, but now mainly with man-made fibres) to give better cohesion and easier application to surfaces with low or variable suction (e.g. building boards or metal lathing). There are also special purpose plasters, such as *low-setting expansion* types (e.g. *board plasters* for use on ceiling or wall boards) and *concrete bonding* types. Some types are specially retarded to be suitable for spray application. There are also *premixed* plasters, containing lightweight or special aggregates. These include plasters which are used for thermal and acoustic plasterwork, anti-condensation and fire-resisting surfaces, and superior bond (especially exfoliated vermiculite), workability (ex-

B

panded perlite) and resilience. Premixed plasters may additionally contain a plasticiser.

PLASTER CLASSIFICATION AND TESTS

BS 1191 'Gypsum Building Plasters' is in two parts. Part 1 lists and deals with four classes of plaster: Class A—*plaster of Paris*, Class B—*retarded hemihydrate*, Class C—*anhydrous* and Class D—*Keene's*. Part 2 deals only with *premixed lightweight* plasters consisting of lightweight aggregate and retarded hemihydrate gypsum plaster.

The properties of Classes A, B and C plasters have already been described. Class D plaster is an anhydrous form made from selected high purity gypsum calcined at a higher temperature than Class C. They are easily brought to a smooth, intensely hard finish and have a slow rate of setting. Anhydrite plasters are not referred to in BS 1191 since they are not currently produced in Great Britain.

Works tests

Routine control tests made at the works usually include the determination of *setting time* using a Vicat apparatus, and *specific surface* by flowmeter. In addition, tests to BS 1191 are made as required.

BS tests for Classes A, B, C and D plasters

These tests are made for chemical composition and freedom from coarse particles (excluding Class B undercoat plasters); soundness, transverse strength (Class A and Class B undercoat and dual-purpose types); mechanical resistance (Class B finish and dual-purpose, and Classes C and D finish plaster); expansion on setting (Class B, board finish only). Table 2, page 302, gives the BS requirements for tests (except chemical composition).

1 *Chemical composition*
This test is to ensure adequate purity and freedom from harmful substance. The content, by chemical analysis, of sulphur trioxide, calcium oxide, soluble sodium and magnesium salts is determined, as well as the loss on ignition, and for metal lathing plaster (Class B) the free lime content.

2 *Freedom from coarse particles*
Unduly coarse particles would prevent a good finish and would not

hydrate readily. Maximum limits are specified for the percentage of material retained on a BS 14 sieve (1·20 mm mesh size).

3 Soundness

'Pats' of neat plaster, cast in ring moulds 4 in (102 mm) diameter and ¼ in (6·4 mm) deep and stored in a humidity cabinet, are steamed at atmospheric pressure for 3 hours and then examined for signs of disintegration, popping or pitting.

4 Transverse strength

Cast prisms, 4 in (102 mm) long and 1 in (25·4 mm) square, of mix proportions which are determined by the class of plaster which has been used to make them are tested for transverse strength (i.e. as beams) after specified storage, with central loading and supports at 3 in (76·2 mm) centres (Fig 5). The breaking load is expressed as a modulus of rupture.

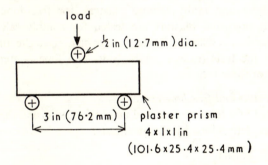

Fig 5 Plaster transverse strength test

5 Mechanical resistance

A steel ball-bearing, ½ in (12.7 mm) diameter and weighing 8·33 g is dropped 6 ft (1·82 m) onto the surface of cast specimens, (dimensions shown in Fig 5). The specimens are of neat plaster mixed to a defined consistency, and are oven-dried before they are tested following storage under prescribed conditions. The surface indentation is a measure of hardness.

6 Expansion on setting

A special *extensometer* (Fig 6) is used to measure the expansion of a specimen made from plaster of standard consistency.

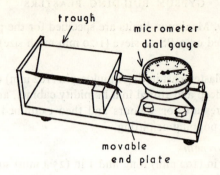

trough

micrometer
dial gauge

movable
end plate

Fig 6 Plaster extensometer

BS tests for premixed lightweight plasters

Tests for undercoat premixed plasters are for soluble salts content, dry bulk density, dry set density, compressive strength and, for metal lathing and multi-purpose plasters, for free lime content. Final coat premixed plasters are tested for soluble salts content and mechanical resistance. Table 3, page 303, gives the BS requirements for the tests (except soluble salts and lime content) and a summary of these follows:

1 Soluble salts and free lime content
The content of soluble sodium and magnesium salts and the free lime content are determined by chemical tests.

2 Dry bulk density
The plaster (powder) is poured without compaction into a 6 in (152 mm) cube container, struck level with the top edge, and the amount of plaster weighed to find the bulk density.

3 Compressive strength and dry set density
Specimens cast to the size of 1 in (25·4 mm) cubes from plaster mixed to a specified consistency are weighed and tested in compression after damp storage for 24 hours, followed by oven-drying to constant weight at (35–40°C).

4 Mechanical resistance
The test follows the procedure of that given for the mechanical resistance of Classes B, C and D final coat plasters, except that 1 in (25·4 mm) cube specimens are used.

EXPERIMENT 6   Setting time of a gypsum plaster

Note The setting time of plaster varies with the water content. This could be investigated by making a number of tests using the method of this experiment

Apparatus Vicat apparatus as for Experiment 1 (see Fig 3), fitted with needle (1 mm² section)—palette knife and mixing board—measuring cylinder (100 ml³)—stopclock—plastics throwaway cup (100 ml³)—physical balance (metric)

Specimen Retarded hemihydrate gypsum plaster

Method
1 Weigh out 200 g of plaster onto the mixing board
2 Fill the measuring cylinder with water
3 Start the stopclock and mix water with the plaster to give a plastering consistency—*note the amount used*
4 Fill the mix into the plastics cup and place it on the Vicat apparatus
5 Lower the needle to touch the plaster surface and release it
6 Re-check with the needle (Section 5) at intervals, at different points on the surface, until there is no penetration. *Note the time*

Result The setting time is the total time from first adding water. Record also the amount of water used as a percentage of the weight of plaster

Note This experiment gives an approximate result for the final setting time of a plaster for a given water content. If desired, more precise data could be obtained using two Vicat needles to obtain an initial and final set respectively, as defined for cements in Experiment 2.

EXPERIMENT 7   Plaster soundness (*pat*) test

Apparatus Six metal ring moulds, 4 in (102 mm) diameter, ¼ in (6·4 mm) deep, with baseplate—beaker (250 ml)—palette knife and mixing board—damp closet—steam chamber—physical balance (metric)—vaseline (petroleum jelly)

Specimen Retarded hemihydrate or anhydrous plaster

Method
1 Smear vaseline on mould surfaces

2 Mix 400 g of plaster with water to give a stiff, plastic paste
3 Fill the moulds with paste, using a pressing action, and smooth the surface level
4 Leave the pats to set in air of relative humidity at least 80% for:
[a] 16 to 24 hours for hemihydrate plasters
[b] 3 days for anhydrous plasters
5 Steam the pats in moulds for 3 hours (keeping them clear of falling drops) using saturated steam at atmospheric pressure
6 Examine pats for disintegration, popping or pitting

Results Record your observations (Section 6) and see Table 2, page 302.

EXPERIMENT 8 Transverse strength of a gypsum undercoat plaster

Note Details of the dropping ball penetrometer (Fig 7) for measuring plaster consistency are given in BS 1191. This is a device for dropping a methylmethacrylate ball of 1 in (25·4 mm) diameter (weighing 10·0–10·5 g) from a height of 10 in (254 mm) onto a surface of freshly mixed plaster filling a ring mould (80 mm internal diameter and 40 mm height), together with a bridge to measure penetration. Alternative means of doing this can be improvised

Apparatus Transverse testing machine—ventilated oven (35–40°C) —dropping ball penetrometer, with ring mould and baseplate—gang mould to cast six specimens, each 4 × 1 × 1 in (102 × 25·4 × 25·4 mm) (Fig 8)—mixing board and palette knife—brass tamping rod, $\frac{1}{4}$ in (6 mm) square—measuring cylinder (250 ml)—damp closet—physical balance (metric)—stopclock—vaseline

Materials for specimens Retarded hemihydrate undercoat plaster, previously stabilised by exposure, in a $\frac{1}{4}$ in (6 mm) layer, to room air for 3 days. Washed sand passing BS 18 (850 μm) sieve, with not more than 10% passing BS 25 (600 μm) (Leighton Buzzard sand is specified for BS tests)

Method

[a] *To prepare a mix of standard undercoat consistency*, of proportions 1:3 by weight plaster: sand
1 Mix together, dry, 200 g of plaster and 600 g of sand
2 Mix with 120 ml of water (as a first trial) to the following schedule: Add the dry sanded mix to water over a period of 30 seconds and

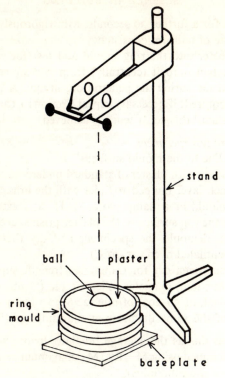

Fig 7 Dropping ball penetrometer

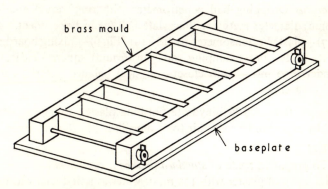

Fig 8 Six-gang mould for 4 in x 1 in x 1 in prisms

allow to soak for a further 30 seconds. Mix vigorously for 1 minute to give a paste of uniform consistency

3 Fill the plaster into the ring mould and test for consistency by dropping the ball onto it (centrally) from a clear height of 10 in (254 mm) and measuring the penetration, in mm. A penetration of 9–10 mm is required: if necessary repeat Sections 1 to 3 using more, or less, water until this consistency is achieved

[b] *To prepare test specimens*
1 Apply vaseline to the mould surfaces
2 Fill each mould with plaster of standard undercoat consistency, in two layers, each layer tamped 10 times with the brass rod
3 Store the mould in a damp closet (R.H. at least 90%) for 24 hours, scrape the top surface of the plaster prisms level with the top of the mould, demould the specimens and dry them to constant weight in a ventilated oven (35–40°C)
4 Test each specimen dry for transverse strength, supported on its side by $\frac{1}{2}$ in (12·7 mm) diameter rollers at 3 in (76 mm) centres, with the load applied, at 50–200 lbf/min (22·6–90·7 kgf/min), through a third roller of the same diameter at the mid-span (Fig 5)

Result Express this as the average breaking load, and modulus of rupture (R) obtained using $M = Rz$ (M = maximum bending moment at failure, z = section modulus).

EXPERIMENT 9 Setting expansion of a gypsum final coat plaster

Apparatus Dropping ball penetrometer with ring mould and 1 in (25 mm) diameter methylmethacrylate ball (as in Experiment 8 and Fig 7)—plaster extensometer to BS 1191 (Fig 6)—mixing board and palette knife—brass tamping rod $\frac{1}{4}$ in (6 mm) square—measuring cylinder (250 ml)—damp closet—physical balance (metric)—stop-clock—vaseline

Specimen Retarded hemihydrate finishing plaster, previously stabilised by exposure to room air for 3 days in a $\frac{1}{4}$ in (6 mm) layer

Method
[a] To prepare a paste of *standard final coat consistency*
1 Mix 350 g of plaster with 120 ml of water as a first trial within 30 seconds

2 Allow to soak for a further 30 seconds, then mix vigorously for 1 minute

3 Fill the mixed plaster into the penetrometer mould and strike it off level with the rim

4 Check the penetration of the ball, in mm, for a clear drop of 10 in (254 mm)

Repeat Sections 1 to 3, but using more, or less, water as necessary to obtain a penetration between 15 and 16 mm (denoting a paste of standard final coat consistency)

[b] Setting expansion

1 Apply vaseline to the trough surfaces, line it with thin, non-absorbent, glazed paper, and check for free movement of the dial gauge stem

2 Hold shut the movable end-plate of the trough and fill it with plaster

3 Ease the movable end-plate horizontally so that it just clears the trough walls (this prevents sticking), then press the plaster to bring it back into firm contact with the plate

4 Place the filled extensometer in a damp closet (at least 90% R.H.) and note the reading (*initial*) of the dial gauge (or adjust it to zero)

5 After 24 hours check the dial gauge reading (*final*)

Result Record the expansion, if any, after 24 hours (final minus initial reading) and express it as a percentage of the length of the specimen (trough length).

Calculate also the percentage water content by weight of plaster for the standard final coat consistency.

EXPERIMENT 10 Mechanical resistance of gypsum final coat plaster by the dropping ball test

Apparatus Ventilated oven (25–40°C)—damp closet—metal tube 68 in (1·72 m) long, ⅝ in (16 mm) internal diameter, with stand and clamps—steel ball-bearing, ½ in (12·7 mm) diameter, weight 8·33 g—gang mould to cast four specimens, each 4 × 1 × 1 in (102 × 25·4 × 25·4 mm)—mixing board and palette knife—measuring cylinder (250 ml)—physical balance (metric)—stopclock—vaseline

Specimen Retarded hemihydrate finishing plaster, previously stabilised by exposure to room air in a thin ¼ in (6 mm) layer for 3 days.

Preliminary work The standard final coat consistency of the plaster must be known, or should be determined, as in Experiment 9, part [a]

Method

1 Smear vaseline on mould surfaces

2 Add 400 g of plaster, within 30 seconds, to a measured amount of water to give the *standard final coat consistency*

3 Allow to soak for 1 minute, then mix vigorously for 2 minutes

4 Fill plaster separately into each section of the gang mould in two layers, tamping each layer 10 times with the brass rod. Strike off the surface level with the top of the mould and place it in a damp closet (at least 90% R.H.) for 24 hours

5 Demould specimens at 24 hours and dry them to constant weight in a ventilated oven (35–40°C) before testing

6 Clamp the tube vertically with its lower end 4 in (100 mm) above the specimen (laid on one side as cast)

7 Drop the steel ball through the tube so that it falls 6 ft (1·82 m) before striking the specimen. Obtain one impression on each *side* of each of *four* specimens (8 impressions), located within an area $\frac{1}{4}$ in (6 mm) from the longer centre line and excluding $\frac{1}{2}$ in (13 mm) at each end

8 Measure the diameter of each impression in two directions at right angles

Result Take the mean of the 16 measured diameters.

Summary

Gypsum plasters are made by heating gypsum rock, calcium sulphate dihydrate ($CaSO_4.2\ H_2O$), to produce either the hemihydrate, Class B (with added retarder) or the anhydrous forms, Classes C and D (with added accelerator).

The plasters set on the addition of water (hydration) to form the dihydrate, and give off heat in the process.

Anhydrous plasters set more gradually than hemihydrate plasters, and give a denser material with a harder finish. They may show a double set and can be retempered after the first set. Do not retemper hemihydrate plasters.

Gypsum plasters normally exhibit a setting expansion (delayed expansion following incomplete hydration can cause bond failure).

Plasters of low setting expansion are obtainable.

Plasters are classified as undercoat or browning, finishing and dual purpose types.

There are special grades of plaster for different types of surface (e.g. board, metal lathing and concrete bonding) in addition to premixed plasters (e.g. vermiculited and perlited).

BS tests for normal plasters are for chemical composition, freedom from coarse particles, soundness, transverse strength, mechanical resistance and expansion on setting. Tests for premixed lightweight plasters are for soluble salts and free lime content, dry bulk density, dry set density, compressive strength and mechanical resistance.

3

LIMES

Building limes are widely used, often with cement or plaster, for such purposes as brickwork mortars, rendering and plastering. They are obtained mainly by the burning of limestone or chalk, and, to a lesser extent, as an industrial by-product. Relatively pure calcium carbonate yields a non-hydraulic *high calcium* (white) lime, but where a significant amount of clay is present in the raw material, a *semi-hydraulic* (grey) lime is obtained. The term 'hydraulic' was given to denote the ability of these limes to set underwater—that is in the absence of air or carbon dioxide, and this results from the process of hydration. An *eminently hydraulic lime* is a lime whose composition and properties approach those of a Portland cement, such as lime made from limestone of the *Blue Lias* formation. Such limes, sometimes called *natural* or *Roman cements,* are not produced commercially to any extent in Great Britain today, since the composition of the raw material makes it suitable for the production of a Portland cement.

The calcination of a pure chalk or limestone gives the following reaction:

Calcium carbonate + heat = quicklime + carbon dioxide (gas)
$$Ca\ CO_3 \qquad = \quad CaO \ + \qquad CO_2$$

In some cases magnesium carbonate is also present, which will give a reaction:

Magnesium carbonate + heat = magnesia + carbon dioxide (gas)
$$MgCO_3 \qquad = \quad MgO \ + \qquad CO_2$$

Where more than 5% of magnesium oxide is present in the finished lime it is classified as a *magnesian lime.* Most of these are from

44

dolomitic limestone and have more than 30% of magnesium oxide.

Where no materials other than the two carbonates mentioned are present, the quicklime, or magnesia, is wholly free to combine with water to form *hydrated lime*. This process is known as *slaking*. Limes with hydraulic properties will have part of their lime content in chemical combination, with only the uncombined or *free lime* available for slaking.

The slaking of quicklime, when fresh, is a very vigorous reaction in which great heat is produced, and the material undergoes considerable expansion:

Quicklime+water=hydrated lime (calcium hydroxide)

$$CaO \quad + H_2O = \qquad Ca(OH)_2$$

A corresponding reaction occurs with magnesia, but only slowly:

Magnesia+water=hydrated lime (magnesium hydroxide)

$$MgO \quad + H_2O = \qquad Mg(OH)_2$$

All limes must be fully slaked before use in building work, otherwise they will be liable to slake at a later date, and cause instability by expansion. Whereas the conversion of calcium carbonate by calcination takes place at a temperature of about 900°C, the corresponding temperature in the case of magnesium carbonate is much lower (about 500°C) and it follows that if both are present the magnesium oxide is likely to be overburnt. Such overburnt material will not hydrate readily and the slaking must be carefully controlled.

Slaking used invariably to be done on site and the entire process took several weeks. But it is much more easily and efficiently done under the controlled conditions of a works, and this is now the usual procedure. This applies to both high calcium and semi-hydraulic varieties, and in the latter case just sufficient water is added to slake the free lime, since excess water would cause a reaction with the hydraulic constituents and reduce the available strength.

Lime in the unslaked form is commonly termed *lump lime*, after the form in which it is delivered. Hydrated lime is delivered in bags in powder form.

When mixed with water as a putty (or 'milk' of lime), pure hydrated high calcium lime undergoes no chemical setting action. Any early stiffening which occurred would be due to loss of water by evaporation or by suction into an adjacent material. For this reason high calcium limes are gauged with cement, plaster or other binder for normal uses in building. Such limes do undergo a very slow hardening due to chemical change, by absorption of carbon dioxide from the atmosphere. This process is known as *carbonation*:

Hydrated lime+carbon dioxide=calcium carbonate+water

$$Ca(OH)_2 \quad + \quad CO_2 \quad = \quad CaCO_3 \quad + H_2O$$

The action progresses only slowly from the surface of the material inwards and since it involves slight shrinkage can give rise to surface crazing or cracking. The formation of water in this reaction explains why lime-plastered surfaces are not usually papered or given impermeable decoration until some months after plastering.

The *storage* of limes must take account of their type. Quicklime cannot be stored satisfactorily for any length of time except in airtight containers, owing to the slaking tendency and the accompanying effects. Limes with hydraulic properties must receive the same consideration in storage as Portland cements.

Hydrated high calcium limes are not affected adversely by moisture, but will very gradually carbonate, resulting in hard 'air-set' lumps forming at the outer layers in a container not air-tight. These limes are often mixed with water (*lime putty*) several days before use, since this gives improved workability.

CLASSIFICATION OF LIMES

BS tests and requirements

BS 890 'Specification for building limes' deals with quicklimes and hydrated limes (powder and putty forms) of both the high calcium and semi-hydraulic types (including magnesian limes). It does not deal with eminently hydraulic limes.

Requirements are specified for chemical composition, fineness and workability (all types), soundness (hydrated limes) and hydraulic strength (semi-hydraulic types). In addition, for lime supplied in putty form, there are tests for density and water loss on standing. A summary of the procedure of these tests follows, and the BS test limits for hydrated lime powder and putty are given in Table 4 on page 304.

1 *Chemical composition*

All types are subject to limits for the analysed carbon dioxide content, insoluble matter, lime and magnesia, and, for semi-hydraulic types, the soluble silica content. High calcium by-product limes are additionally tested for soluble salts content.

2 *Fineness*

Maximum limits are given for the residue after washing a sample through specified fine-mesh sieves.

3 *Soundness*

[*a*] *General expansion test*—the test is similar to that for the soundness of Portland cement using the Le Chatelier moulds, except that lime specimens are steamed for 3 hours, instead of being boiled. The mix used is 1 part of the lime sample to 4 parts of a standard sand and $\frac{1}{3}$ part of ordinary Portland cement, together with $\frac{2}{3}$ part of water. The proportions are measured by weight.

[*b*] *Pat test*—a sample of a standard lime putty is prepared and a proportion of plaster of Paris is added. Pats, cast in ring moulds, are oven-dried, steamed for 3 hours and then examined for signs of disintegration, popping or pitting.

4 *Workability*

A special apparatus, the standard *flow table* (Fig 9) is used. Details of this apparatus are given in BS 890. The platform of this table can be

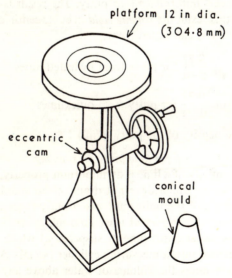

Fig 9 Standard flow table

alternately raised slowly and dropped suddenly, through a height of 12·7 mm by the action of an eccentric cam. Lime putty prepared to a standard consistency is filled into a metal cone at the centre of the table. The workability is given by the number of bumps for the demoulded putty to spread to a circle of diameter 190 mm.

5 Hydraulic strength

4:1 sand-lime mortar specimens, made using a standard sand are cast in moulds 4 in (100 mm) long and 1 in (25·4 mm) square. After storage under specified conditions for 28 days the specimens are immersed in water for 4 hours then tested wet for transverse strength (as beams), supported on a cast side at 3 in (76 mm) centres and centrally loaded. The result is expressed as the modulus of rupture.

6 Density of lime putty

In addition to determining compliance with the BS limit for density of the lime putty as received, this test can be applied to other limes to determine their putty volume yield at standard consistency.

The putty density is determined by weighing a cylindrical vessel (60 mm diameter and 90 mm height, internally) before and after filling it with the uniformly mixed putty. The result is expressed in g/cm³. From this result the *volume yield* in cubic centimetres of putty per gram of lime is calculated from:

$$\frac{0{\cdot}57}{d-1} \text{ for putty from hydrated limes}$$

$$\text{or } \frac{0{\cdot}72}{d-1} \text{ for putty from quicklimes}$$

where d is the density of the putty (g/ml)

7 Separation of water from lime putty on standing

The water-retentivity of a lime is an important property, and for test purposes the water which does not drain off on standing must be known. A sample of the uniformly mixed putty (1500 g) is levelled inside a cylinder of capacity at least 1200 ml, covered and left for 24 hours. Any water (with lime in suspension) which rises to the surface is decanted into a measuring cylinder (50 ml). After this has settled for 30 minutes the volume of water above any sediment is recorded.

EXPERIMENT 11 Lime soundness (pat) test

Apparatus Mixing board and palette knife—measuring cylinder (100 ml)—beaker (250 ml)—ring moulds, 100 mm internal diameter, 5 mm deep, with internal taper 5°, including baseplate—steam chamber—ventilated oven (40 ± 5°C)—physical balance (metric)—vaseline.

Materials Building lime (powdered hydrate). Plaster of Paris ($2CaSO_4 . H_2O$ unretarded)

Method
Prepare separately 3 pats, as follows:
1 Mix 70 g of lime with 70 g of water and leave 2 hours to soak
2 Apply vaseline thinly to mould surfaces
3 Remix the putty, adding water if necessary to make it plastic
4 Add 10 g of plaster of Paris and remix for 2 minutes
5 Set the ring mould with its wider diameter against the baseplate and fill it in small quantities, pressing with the palette knife to exclude air, and finally smooth the surface level with not more than 12 strokes of the blade.
Note Sections 4 and 5 are to be completed within 5 minutes
6 Allow each pat to set for 30 minutes then dry it overnight in a ventilated oven at 40°C, removing the ring mould, if desired, but not the baseplate
7 Reject any pats with shrinkage cracks and replace them with new pats. Now steam the pats on baseplates at atmospheric pressure for 3 hours, and protect them from falling drops
8 Examine the pats for any disintegration, popping or pitting.

EXPERIMENT 12 Workability test for limes

Note The preparatory work, section [*a*] below, must be started 24 hours before making the workability test

Apparatus BS standard flow table (Fig 9) fixed to a rigid base, and conical mould—mixing board—gauging trowels—palette knife—measuring cylinder (500 ml)—physical balance (metric)—blotting paper

Specimen Building lime (powdered hydrate)

Method
[*a*] To prepare putty of *standard consistency*

1 Mix 500 g of lime and 500 g of water to a uniform paste, cover to prevent evaporation and stand it aside 24 hours

2 Remix the steeped lime, using the trowels, for 3 minutes to produce a uniform putty

3 Rinse the conical mould with water, drain and shake it, then place it on the flow table at the centre of the platform with the neck uppermost

4 Fill the mould with putty, level the surface and lift off the mould

5 Operate the flow table to give one bump and measure the diameter of the putty (average for 3 directions). This must be 110 mm ± 1 mm for standard consistency

6 *If necessary*, adjust the consistency of the putty by adding water, or extracting it (using blotting paper), each time remixing the putty for 2 minutes

Note The preparation of the putty (Sections 2–6) must be completed within 11 minutes. (If it is not, discard it for a fresh sample.) The workability test should follow immediately

[*b*] *Workability test*

7 Determine the total number of bumps for the lime putty of standard consistency to reach a spread of 190 mm ± 1 mm (average of 3 diameters), at a steady rate of 1 bump per second.

Summary

Building limes are made by calcining chalk or limestone and then slaking the resulting quicklime to give hydrated lime. Incomplete slaking would constitute unsoundness in lime, which could cause expansion failures in the work.

A magnesian lime is defined as one containing more than 5% of magnesium oxide (from magnesian limestone or dolomite). White (high calcium) limes have no setting action with water, but will carbonate slowly due to atmospheric carbon dioxide. These limes are used principally as a workability aid in mortars, plastering and rendering (also in some lightweight concretes and in sandlime bricks).

Grey limes have hydraulic properties—they set and harden due to chemical action with water (hydration).

Tests for limes given in BS 890 include chemical composition, fineness and workability, soundness of hydrated limes and strength of hydraulic limes. Lime putties are tested for density and loss of water on standing.

4

AGGREGATES

An aggregate is a material in granular or particle form, such as sand or gravel, which is added to the class of materials known as *binders* (e.g. cements, hydraulic limes, plasters and bitumen) to produce a solid mass on hardening. Since most aggregates are inert and undergo no chemical action with the binder, the strength of the combined mass depends on

[a] the specific adhesion or *bond* which develops between aggregate and binder
[b] the mechanical *key* or interlock which develops between the constituent particles in virtue of their shape, size and surface texture
[c] the strengths of the aggregate and binder respectively

Reasons for mixing an aggregate with a binder may include one or more of the following:

[a] to reduce material costs, using the aggregate as an extender (where the aggregate is less costly than the binder)
[b] to offset either the drying shrinkage or setting expansion of a binder
[c] to obtain increased, or reduced, density (by using high or low density aggregates respectively)
[d] to alter appearance, such as colour or texture
[e] to obtain better resistance to wear by abrasion or weathering (using hard, abrasion resistant or non-absorbent aggregates)
[f] to impart some other special property such as fire resistance, thermal insulation or acoustic characteristics
Aggregate may be classified as fine or coarse, natural or artificial, dense or lightweight and so on, according to the requirements of the user.

The descriptions *fine* or *coarse* refer to the largest size of particle present in substantial amounts. For general building purposes, a *fine* aggregate is one which will mainly pass a $\frac{3}{16}$ in (4·76 mm) square-mesh sieve, and a *coarse* aggregate is one which will be mainly retained on that sieve. An aggregate which has substantial amounts of both fine and coarse particles is termed an *all-in* aggregate.

Aggregates from *natural sources* include crushed or uncrushed stone or gravel, and sand, including crushed stone sand and crushed gravel sand. The crushing is done by machinery, usually at the pit or quarry. *Artificial* aggregates include crushed brick, blast-furnace slag and numerous lightweight and special aggregates.

Fine aggregates are used, together with appropriate binder, to produce material for such purposes as rendering, plastering, floor topping, and road surfacing. In addition to this, aggregates are used with a cement binder as jointing material for pipes, bedding for tiles, and mortars for brickwork and blockwork. For concrete, fine aggregates are normally used together with cement and coarse aggregate or, alternatively, all-in aggregate can be used. Coarse aggregate is also included in some grades of asphalt.

LIGHTWEIGHT AGGREGATES

Lightweight aggregates are composed of particles of high porosity, resulting in a low bulk density. The use of lightweight aggregates reduces the dead-weight of a structure, allowing the use of smaller supporting members and foundations at reduced cost. It also gives improved thermal insulation.

Clinker

Clinker should be well-burnt *sintered* or *fused* furnace residue, substantially free of such impurities as sulphides and quicklime, which promote unsoundness (expansive chemical action). Coke breeze is therefore not to be classified as a clinker, as it is unsintered and may contain harmful impurities.

Foamed blast-furnace slag

This is crushed and screened expanded slag, produced by water-cooling the molten slag obtained in the production of pig-iron.

The expansion results from the bloating action of the resulting steam. The material is to be distinguished from the heavier *air-cooled slag*, which is also used as an aggregate.

Pulverised fuel ash (PFA)

This is the ash of pulverised fuel (powdered coal), recovered from the flue gases of power stations. Its normal form is that of an extremely fine powder, but it is also obtainable in a processed, nodular form (*sintered PFA*).

Expanded clay, shale, slate, perlite and vermiculite

These are expanded forms of the natural material produced by heating. *Perlite* is a glassy volcanic rock, giving expanded perlite, and *vermiculite* is a mineral of plate-like structure, like mica, giving exfoliated vermiculite.

Pumice

Pumice is a natural, highly porous rock of volcanic origin.

Diatomaceous earths

Diatomaceous earths are naturally occurring siliceous deposits formed by the decomposition of microscopic plant skeletons (algae).

Sawdust and wood fibre

These can normally be used as aggregates only after *stabilisation* (chemical treatment) to prevent reaction between substances present and the cement or other binder, and to control subsequent moisture movement.

HEAVY AGGREGATES

Heavy aggregates are used mainly for the special purpose of screening radioactivity, an important consideration with the advent of nuclear power. Aggregates suitable are those with specific gravities greater than about four. They include certain minerals, notably *barytes* (barium sulphate) s.g. 4·6, and *haematite* (an iron oxide) s.g. 5·0, in addition to the metals iron and steel.

The classification of an aggregate as *fine* or *coarse* would give little chance of ensuring its suitability for a particular purpose. For example, a sand suitable for concreting work may be far too coarse to have any value as a material for use in plastering final coats. In practice we are concerned not only with the maximum particle size of an aggregate, but with the amounts of the various sizes present. This is known as the *grading*. The grading of an aggregate is found by shaking a sample through a series of sieves (Fig 10) of different standard mesh size. The amount retained on each sieve is weighed and expressed as a percentage of the sample. It is conventional to convert these results to give the percentage weight of the whole sample to pass each sieve (see Experiment 13 and the example in Fig 13).

Grading requirements depend on the proposed use of the aggregate and in this book they are referred to in the various sections which deal with the types of work concerned.

Normal sizes:
medium mesh 12 in dia.(305 mm)
fine mesh 8 in dia.(203 mm)

Fig 10 Test sieve for aggregates

Sampling aggregates

It is important that any sample of aggregate used for test purposes is truly representative of the bulk from which it is taken. This can usually be ensured by taking the sample in a number of increments (10 is common) from various positions and levels in a stockpile, or bin, and mixing them together. If this provides more material than is needed, the sample may then be reduced by *quartering* (Fig 11); in other words by dividing it into four similar parts, rejecting two, which are diametrically opposite and remixing the remaining two. The process can be repeated as necessary. An alternative to quartering is the use of a *sample divider*. Fig 12 shows a type known as a *riffle-box*.

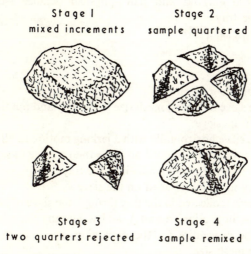

Stage 1
mixed increments

Stage 2
sample quartered

Stage 3
two quarters rejected

Stage 4
sample remixed

Fig 11 Reducing an aggregate sample by quartering

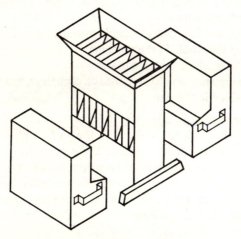

Fig 12 A sample divider (riffle-box)

EXPERIMENT 13 Grading test for fine aggregate (sieve analysis)

Apparatus British Standard sieves (to BS 410), 8 in (200 mm) diameter and to the following mesh sizes: $\frac{3}{16}$ in (4·76 mm), No. 7 (2·40 mm), No. 14 (1·20 mm), No 25 (600 μm), No. 52 (300 μm), No. 100 (150 μm) lid and pan—physical balance (sensitive to 0·2 g)—large tray

Specimen Dry sand

Method
[a] *For hand sieving:*
1 Weigh out 200 g of sand
2 Stand the sieve of largest mesh size in the tray and put the weighed sample onto the sieve
3 Shake the sieve horizontally with a jerking motion in all directions for at least 2 minutes and until no more than a trace of sand passes. Ensure that all sand passing falls into the tray
4 Weigh any material retained on the sieve
5 Pass material collected in the tray through the sieve of next smaller mesh size, as in Sections 2 and 3, and weigh any material retained
6 Repeat the procedure for the remaining sieves, in the order of diminishing mesh size

[b] *For mechanical sieving where a vibrating machine is available*
1 Fit the sieves and pan together to form a 'nest' (mesh size diminishing towards the bottom)
2 Tip the weighed (e.g. 200 g) sample into the nest of sieves, fit the lid and clamp the whole in the machine
3 Operate the machine for 20 minutes
4 Weigh the material which is retained on each sieve, and the material which is retained in the pan

Results Record these in a test schedule, as shown by the example in Fig 13. In the first column of the table the various BS *sieve sizes* are listed. The second column contains the *experimental data*. In the third column, the weights retained have been converted to *percentages* of the total sample weight. The fourth column shows the total weight of the sample *passing* each sieve, as a percentage, calculated from the preceding column as follows:

$$100 \ - \ 0·8 = 99·2$$
$$99·2 - \ 2·5 = 96·7$$

$$96 \cdot 7 - 13 \cdot 7 = 83 \cdot 0$$
$$83 \cdot 0 - 49 \cdot 0 = 34 \cdot 0$$
$$34 \cdot 0 - 30 \cdot 2 = 3 \cdot 8$$

The final column gives the results to the nearest 1%. A grading 'curve' (Fig 14) may be drawn to illustrate the results with straight lines connecting the plotted points. Special graph paper with a horizontal logarithmic scale is often used for this purpose.

BS sieve (mesh size)	Material retained		Material passing (per cent by weight)	
	weight gf	per cent to 0.1%	actual to 0·1%	to nearest 1%
3 in (76·2 mm)				
1½ in (38·1 mm)				
¾ in (19·0 mm)				
⅜ in (9·53 mm)				
³⁄₁₆ in (4·76 mm)	0	0	100·0	100
No. 7 (2·40 mm)	1·6	0·8	99·2	99
No. 14 (1·20 mm)	5·0	2·5	96·7	97
No. 25 (600 μm)	27·4	13·7	83·0	83
No. 52 (300 μm)	98·0	49·0	34·0	34
No. 100 (150 μm)	60·4	30·2	3·8	4
Passing 100	7·6	3·8		
TOTAL	200·0	100·0		

Fig 13 Example of a grading test schedule

EXPERIMENT 14 Grading test for coarse aggregate (sieve analysis)

Apparatus British standard sieves (to BS 410) of 12 in (300 mm) diameter and to the following mesh sizes: 1½ in (38·1 mm), ¾ in (19·05 mm), ⅜ in (9·53 mm)—lid and pan—physical balance (sensitive to at least 2·5 g)—large tray

Specimen Dry coarse aggregate, ¾–³⁄₁₆ in (19–4·75 mm)

Method This is the same as in the preceding experiment (Experiment 13), except for the use of different sieves and a larger sample (see

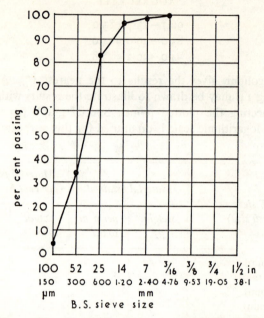

Fig 14 A grading result shown graphically (see Fig 13)

Table 5, page 305). For this test, on ¾ in (19 mm) maximum size aggregate, an amount between 2 kg and 3 kg is appropriate

Results Record these on a test schedule and draw a graph, as for Experiment 13.

Note on the overloading of sieves

In order to avoid overloading and clogging of sieves it may at times be necessary to split a sample into smaller parts, later combining the results for each portion sieved. Table 7, page 306, gives loading limits for different sieves.

Fineness modulus

The grading of an aggregate may be expressed by a single figure, the *fineness modulus*, obtained by adding together the percentage by weight of material *retained* on each of nine test sieves: 1½ in (38·1 mm), ¾ in (19·05 mm), ⅜ in (9·53 mm), 3/16 in (4·76 mm), No. 7 (2·40 mm), No. 14 (1·20 mm), No. 25 (600 μm), No. 52 (300 μm), and No. 100

(150 μm), then dividing by 100. This figure allows a comparison of aggregates for their relative overall fineness, and is useful in routine checks on the grading of successive deliveries of aggregate from a particular source. It has also been used in concrete mix design (e.g. in combining aggregates). Other ranges of sieves can be used, but, of course, results are only comparable for a particular series.

Example 4 Calculate the fineness modulus for a sand with the following grading:

BS sieve	$\frac{3}{16}$ in (4·76 mm)		No. 7 (2·40 mm)		No. 14 (1·20 mm)	
% passing	100		97		91	
Calculation						
% retained	0	+	3	+	9	+

BS sieve	No. 25 (600 μm)		No. 52 (300 μm)		No. 100 (150 μm)	
% passing	63		33		9	
Calculation						
% retained	37	+	67	+	91	= 207

Result Fineness modulus = 207 ÷ 100
$$= 2·07$$

Example 5 Values of fineness modulus for a fine and a coarse aggregate are 2·60 and 6·62 respectively. Calculate [a] the fineness modulus for a blend of 2 parts of the fine to 1 part of the coarse aggregate (by weight) [b] the percentage of each for a blend with fineness modulus 5·0.

[a]   Fine aggregate   2 parts × 2·60 = 5·20
      Coarse aggregate   1 part × 6·62 = 6·62
        Total   3 parts   11·82

Fineness modulus of blend = 11·82 ÷ 3
$$= 3·94 \quad \text{Ans. (a)}$$

[b] If P = percentage by weight of fine aggregate (to total)

$$2 \cdot 60 \text{ P} + 6 \cdot 62 (100 - \text{P}) = 5 \cdot 0 \times 100$$
$$2 \cdot 60 \text{ P} + 662 - 6 \cdot 62 \text{ P} = 500$$
$$4 \cdot 02 \text{ P} = 162$$
$$\text{P} = 162 \div 4.03$$
$$= 40 \cdot 3$$

Giving 40% fine: 60% coarse   Ans. (b)

Alternatively, solve by formula:

$$P = \frac{C_m - T_m}{C_m - F_m} \times 100$$

where C_m = coarse aggregate fineness modulus
F_m = fine aggregate fineness modulus
T_m = type (required) fineness modulus

Combined gradings

Various calculation and graphic methods are available for blending aggregates. They do not all serve exactly the same purpose, and must be selected for use accordingly. You have already seen the *fineness modulus* applied to combining aggregates; it can be used for this purpose in concrete mix design. Attempts made to get a better correlation with specific surface (which itself is not easily determined directly by tests but which is the key factor to an aggregate's influence on workability, see also Chapter 8, 'Concrete') have led to the *surface area index* method. In this the aggregate grading is converted to a single index by allocating *surface area factors* to each size of particle present and summing the products of these factors and the fractions of each size present. The *basic factors* assume particles of *spherical* shape, but can be successfully used in adapted mix design procedures to account for non-spherical particles (the alternative of modifying the factors to fit a mix design procedure has found favour with a number of authors). Example 8 shows the method, using basic factors, but we shall first look at two other examples.

Example 6 Two sands have the gradings shown below. Determine the proportions of each to give a blend with 45% passing the No. 25 (600 micron) sieve. (*Note* The following methods apply equally where sieves other than the stated sizes are used.)

sieve size	$\frac{3}{16}$ in (4·76 mm)	No. 7 (2·40 mm)	No. 14 (1·20 mm)
% passing			
Fine sand (F)	100	98	92
Coarse sand (C)	95	76	42

sieve size	No. 25 (600 μm)	No. 52 (300 μm)	No. 100 (150 μm)
% passing			
Fine sand (F)	84	20	12
Coarse sand (C)	28	14	5

The critical figures for this example are the amounts passing the No. 25 (600 micron) sieve (84% and 28%)

[a] *By calculation*

$$\text{If } P = \% \text{ of fine sand (to total)}$$
$$84\,P + 28\,(100 - P) = 45 \times 100$$
$$84\,P + 2800 - 28P = 4500$$
$$56\,P = 1700$$
$$P = \frac{1700}{56}$$
$$= 30.4, \text{ say, } 30\%$$

Alternatively solve by formula:

$$P = \frac{T-C}{F-C} \times 100$$
$$= \frac{45-28}{84-28} \times 100$$
$$= \frac{1700}{56}$$
$$= 30·4\%$$

[b] *Graphically* (Fig 15) The percentages passing the No. 25 (600 micron) sieve are plotted vertically at the left and right sides of the graph (fine and coarse respectively). These two points are joined by a straight line, *ab*. A horizontal line *XX* is drawn at the 45% pass level required. Vertical line *YY* is drawn through the intersection of *ab* and *XX*. Line *YY* gives the percentage of fine aggregate required (30%), scaling from the right-hand side of the graph.

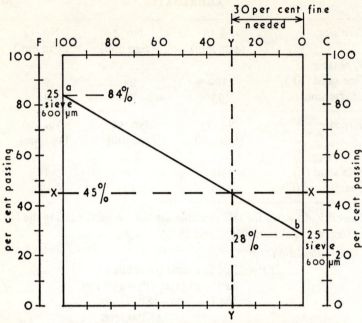

Fig 15 Combining two aggregates (Example 6)

Example 7 From the gradings given below for a fine and a coarse aggregate (F and C), find the combined grading which results from a blend of 30% fine to 70% coarse aggregate. (*Note* The following methods apply equally where sieves other than the stated sizes are used.)

BS sieve size	¾ in (19·05 mm)	⅜ in (9·53 mm)	3/16 in (4·76 mm)	No. 7 (2·40 mm)
% passing F	*	*	100	90
C	100	48	5	0
(a) Calculation				
30%×F	30·0	30·0	30·0	27·0
70% × C	70·0	33·6	3·5	0
Combined	100·0	63·6	33·5	27·0
to nearest				
1%	100	64	34	27

* These must be 100

BS sieve size	No. 14 (1·20 mm)	No. 25 (600 μm)	No. 52 (300 μm)	No. 100 (150 μm)
% passing F	81	66	32	5
C				
	———	———	———	———
(a) Calculation (cont.)				
30%×F	24·3	19·8	9·6	1·5
70%×C				
	———	———	———	———
Combined to nearest	24·3	19·8	9·6	1·5
1%	24	20	10	2

Summary of method Take 30% of grading F and 70% of grading C on all figures, add together for each sieve size, and take the nearest whole number.

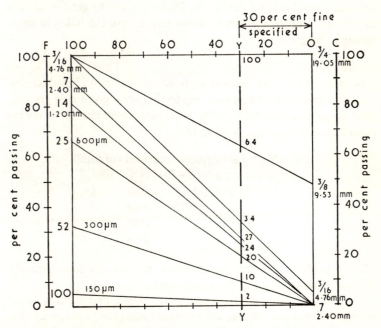

Fig 16 Overall grading by a graphic method (Example 7)

[b] Graphical method

This is shown in Fig 16 and is a variation on the method used for Example 6 (Fig 15). However, this time you must *start* by putting in a *number* of lines such as *ab* (one for each sieve size). Next draw in a vertical line, *YY*, for the specified 30% fine content, and where this line intersects the sloping lines, read off the percentage amounts (shown) which pass the relevant sieves.

Summary of graphical method for surface area index (Example 8)

First convert the grading from cumulative percentages *passing* each sieve to the percentage *retained between* each pair of sieves (these should total 100). Allocate surface area factors, starting with 1 for the largest particle size and doubling for each successive smaller size (1, 2, 4, 8, 16, etc.) This system applies only where the sieve aperture sizes halve successively.

Multiply *percentage retained* figures by the corresponding *surface area factor* and add up the result. Divide by 100 to get the surface area index. The next example shows how to use the index in combining aggregates.

Example 9 A fine and a coarse aggregate are to be combined to give a surface area index approximating to that for a given *type grading*. The known surface area indices are: fine aggregate 49·9, coarse aggregate 2·90 and type grading 18·9.

If P = % of fine aggregate required (to total)

$$49·9\ P + 2·9\ (100 - P) = 18·9 \times 100$$
$$49·9\ P + 290 - 2·9\ P = 1890$$
$$47·0\ P = 1600$$
$$P = \frac{1600}{47·0} = 34·0\%$$

The proportions required are therefore 34% of the fine aggregate and 66% (100–34) of the coarse aggregate.

Usually the actual gradings would be known, and a check could be made to see how closely the *combined* and *type* gradings agreed. There are other methods of combining aggregates.

Example 8 Calculate the surface area index for an aggregate with the grading shown below

Sieve size	1½ in (38·1 mm)	¾ in (19·05 mm)	⅜ in (9·53 mm)	3/16 in (4·76 mm)	No. 7 (2·40 mm)	No. 14 (1·20 mm)	No. 25 (600 μm)	No. 52 (300 μm)	No. 100 (150 μm)	Total
% passing	100	96	56	33	28	25	20	10	1	

Calculation:

Particle size	1½–¾	¾–⅜	⅜–3/16	3/16–7	7–14	14–25	25–52	52–100	P/100	Total
% between sieves (P)	4	40	23	5	3	5	10	9	1	100
Surface area factor (A)	1	2	4	8	16	32	64	128	256	—
Surface area product (P×A)	4	80	92	40	48	160	640	1152	256	2472

Surface area product $4 + 80 + 92 + 40 + 48 + 160 + 640 + 1152 + 256 = 2472$

Surface area index $= 2472 \div 100 = 24·72$

c

When measuring sand by volume, allowance should be made for the fact that it can occupy a greater volume when damp than when dry. This effect is known as *bulking*. The extent of the bulking varies with the moisture content of the sand, but is governed by the film thickness of the water around its particles, and the particle sizes (grading).

This is shown by the graph in Fig 17, from which you will see that fine sands can bulk more than coarse sands. The amount of bulking is given by a standard formula:

$$\frac{\text{Bulking}}{(\%)} = \frac{\text{damp (bulked) volume} - \text{dry volume}}{\text{dry volume}} \times 100$$

However, use is made for tests of the fact that sand when inundated by water occupies the same volume as when dry, giving:

$$\frac{\text{Bulking}}{(\%)} = \frac{\text{damp (bulked) volume} - \text{inundated volume}}{\text{inundated volume}} \times 100$$

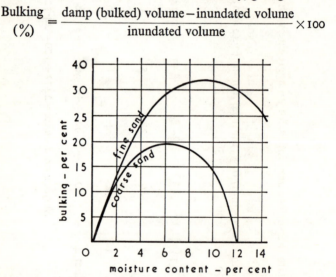

Fig. 17 Typical bulking curves for fine and coarse sands

Bulking test

The test is easily made using a flat-bottomed cylindrical container. Also required are a steel rule, a rod and a tray. The method is illustrated in Fig 18. Damp sand is *loosely* filled into the container to about $\frac{2}{3}$ of its capacity, and the depth, D, is measured. This sand is tipped into a tray. The container is half filled with water and the damp sand returned into it, a little at a time, rodding it to displace air. The depth, d, of this inundated sample is noted. The result is then calculated from:

$$\text{Bulking} \atop (\%) = \frac{D-d}{d} \times 100$$

In practice the result of a bulking test will depend on the size of the container used and the manner of filling it. In consequence where the results are to be applied to the measurement of materials for the works a container of similar size and shape to the gauge box should be used for the test.

An alternative method of finding the bulking of a sample is to calculate it from the known moisture content and values of dry and damp bulk densities (see page 73).

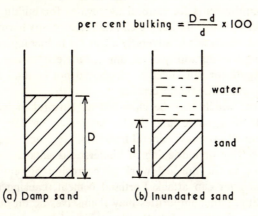

$$\text{per cent bulking} = \frac{D-d}{d} \times 100$$

(a) Damp sand   (b) Inundated sand

Fig. 18 A bulking test for sand

Example 10 Calculate the percentage bulking of a sample of sand to which the following test results apply:

Depth of damp sand in cylinder   200 mm
Depth of same sand inundated   160 mm

$$\text{Bulking} \atop (\%) = \frac{200-160}{160} \times 100 = \frac{4000}{160}$$
$$= 25\%$$

A knowledge of the amount of moisture present in an aggregate may be necessary to allow adjustment of the quantity of water added to a concrete mix so that the specified total is not exceeded. This total is usually given on the basis of dry aggregates. The amount of moisture in an aggregate may also have to be known to allow adjustment in the weight of aggregate measured when weight batching. Various forms of special apparatus are available for the determination of moisture contents. A simple, accurate, *drying method* is given in Chapter 1 of Volume I. For other methods see BS 812.

A moisture content result may be expressed either as a percentage of the dry weight of the sample, or as a percentage of its wet weight.

QUALITY OF AGGREGATES

General requirements for natural aggregates for building are that they should be hard, durable, clean and free of any harmful matter to an extent which would adversely affect the hardening of a binder, or the strength or durability of the finished material in which they are put. Aggregate for reinforced work should not contain substances which will attack the reinforcement.

Substances which may be classified as harmful include certain salts, coal, mica, shale and similar laminated or flaky particles, clay or loam and pyrites (mineral iron sulphide). *Salts* and *clay* may affect the setting or hardening of binders and salts can also cause efflorescence in the finished work or corrosion of reinforcement. *Soluble sulphates* can attack Portland cement, causing expansion and softening. Other impurities may themselves be unsound, causing weakness. *Pyrites* is often the cause of unsightly stains of a rust-red colour on the face of finished work. Excessive *silt, fine dust, loam* or *clay* may prevent proper bond developing between an aggregate and the binder, and may also contribute to the use of an excessive amount of mixing water leading to weakness and porosity in the hardened material. Certain *organic impurities* can retard, or prevent, the setting of cement.

Aggregates which are highly *porous* (for example, some limestones and sandstones) are normally unsuitable for use in external work, or work subjected to severe exposure or chemical attack.

Requirements for the classification and testing of aggregates for various uses in building are given in BS 812 'Methods for sampling and testing of mineral aggregates, sands and fillers'.

Sampling procedures are described, and a system of classification by *geological* group, *petrological* (rock) name, *particle shape* and *surface texture* is detailed.

Tests are included for *particle shape* and *size*, *specific gravity* and *water absorption*, also *mechanical properties* (impact and crushing strength, abrasion and friction values). Tests on the bulk material include *grading* (sieve analysis), *voids* and *density*, *bulking* and *moisture content*, also *silt* and *organic* contents. No limits for test results are given in BS 812, since these are included in the appropriate British Standard (for example, BS 882 for concrete aggregates, BS 1200 for brickwork sands, and so on).

1 *Particle shape*

This is classified as *rounded, irregular, angular, flaky, elongated* or *flaky and elongated*. Typical illustrations are shown in Fig 19.

2 *Surface texture*

This is assessed as *glassy, smooth, granular, rough, crystalline* or *honeycombed*.

3 *Crushing strength*

When assessing the potential value of a *rock for crushed aggregate*, tests are sometimes made on cylindrical specimens cut from the rock. They are crushed in a compression testing machine.

Aggregate in bulk is tested by crushing it in a standard cylinder, with the load applied through a close-fitting plunger (Fig 20). Standard tests give the aggregate crushing value or the 10% fines value.

4 *Aggregate impact value*

This test measures an aggregate's resistance to impact load, as distinct from a gradually applied load. A hammer weighing 30 lbf (14 kgf) drops 15 in (380 mm) onto a weighed amount of oven-dried material (normally $\frac{1}{2}-\frac{3}{8}$ in) (12·7–9·5 mm) packed into a steel

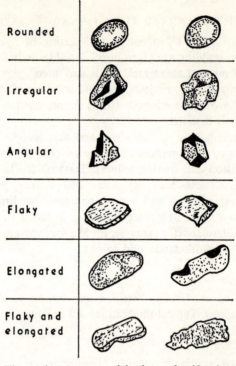

Rounded		
Irregular		
Angular		
Flaky		
Elongated		
Flaky and elongated		

Fig 19 Aggregate particle shape classification

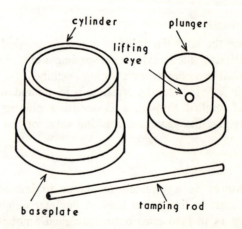

Fig 20 Aggregate crushing strength test-cylinder

cup. After 15 blows the amount crushed finer than a BS 7 (2·40 mm) sieve is determined and expressed as a percentage of the sample weight.

5 Aggregate abrasion value

This test gives a measure of the resistance of aggregates to wear by abrasion. Aggregate particles size $\frac{1}{2}$–$\frac{3}{8}$ in (12·7–9·5 mm), set in a small tray, are abraided by sand in a special machine. At the end of a specified period the loss in weight is determined and expressed as a percentage of the original sample weight.

6 Laboratory determined polished stone value

This test allows a relative assessment of the extent to which different aggregates will polish, and so become slippery under traffic. A layer of particles of $\frac{3}{8}$ in–$\frac{5}{16}$ in (9·5–7·9 mm) material, embedded in cement mortar, is subjected to polishing action by a rotating pneumatic-tyred wheel, under a load of 88 lbf (40 kgf) and with a continuous feed of water and abrasive powder, for six hours. The standard mechanical friction tester has a rubber-faced slider attached to a pendulum which traverses the test surface at the bottom of its swing. The surfaces are tested wet and the height to which the pendulum rises after its traverse is a measure of the surface friction.

7 Flakiness index by thickness gauge

A sample is sieved through standard mesh sizes to obtain fractions retained between different pairs (2$\frac{1}{2}$–2 in/63·5–50·8 mm, 2–1$\frac{1}{2}$ in/ 50·8–38·1 mm, 1$\frac{1}{2}$–1$\frac{1}{4}$ in/38·1–31·75 mm, 1–$\frac{3}{4}$ in/25·4–19·05 mm, $\frac{3}{4}$–$\frac{1}{2}$ in/19·05–12·7 mm, $\frac{1}{2}$–$\frac{3}{8}$ in/12·7–9·53 mm, $\frac{3}{8}$–$\frac{1}{4}$ in/9·53–6·35 mm). Each fraction which constitutes more than 15% of the sample is tested by presenting each of at least 200 pieces to a slot in a metal plate. (The slot width is three-fifths of the appropriate mean sieve size.) For fractions constituting between 5% and 15% of the sample at least 100 pieces are similarly tested. Fractions which constitute less than 5% of the sample are not tested. The weight of material *passing* the various thickness gauges is expressed as a percentage of the total sample weight.

8 *Elongation index by length gauge*

This is the percentage weight of particles in a sample with a length exceeding 1·8 times their mean sieve size (when prepared by sieving and tested in the fractions specified for the flakiness index test). Pieces are checked by presenting them, held lengthwise, to two cylindrical pegs spaced to give the appropriate clearance. The total weight of material *retained* by the various length gauges is expressed as a percentage of the total sample weight.

9 *Angularity number*

This test relates the shape of single sized aggregate particles (e.g. $\frac{3}{4}$–$\frac{1}{2}$ in/19·05–12·7 mm, $\frac{1}{2}$–$\frac{3}{8}$ in/12·7–9·53 mm, $\frac{3}{8}$–$\frac{1}{4}$ in/9·53–6·35 mm or $\frac{1}{4}$–$\frac{3}{16}$ in/6·35–4·76 mm), to spherical particles of the same mean size, by measuring the extent to which they occupy the space inside a cylindrical container (0·1 ft³ or 3 l), into which they are packed.

A result is given by:

$$\text{Angularity number} = 67 - \frac{100 \, W}{C \times G}$$

where W is the weight of aggregate to fill the container (g)
C is the capacity of the cylinder (cm³)
G is the aggregate's specific gravity (oven-dried basis)

10 *Specific gravity, density, voids, bulking, moisture content and water absorption*

These tests relate to properties considered in some detail in Volume 1 (Chapters 1 and 3) with the exception of bulking, which was explained earlier in this chapter. The method of test in BS 812:1967 for the specific gravity of aggregates gives three results—the specific gravity on an oven-dried basis (calculated from the oven-dried weight and the gross, or bulk, particle volume) the specific gravity on a saturated and surface-dried basis (calculated from the saturated surface dry weight and the gross, or bulk, particle volume) and the apparent specific gravity (calculated from the oven-dried weight and the volume excluding the accessible pores, those filled by 24 hours immersion in water).

Different particle sizes of the same material will show different specific gravities whatever the method of expressing a result, and for a coarse aggregate the result should be obtained for *each particle size*; [for example, $1\frac{1}{2}-\frac{3}{4}$ in (38·1–19·05 mm), $\frac{3}{4}-\frac{3}{8}$ in (19·05–9·53 mm), $\frac{3}{8}-\frac{3}{16}$ in (9·53–4·76 mm)] and stated together with the grading, or alternatively the sample *mean specific gravity* may be determined by a single test and stated. The latter is common practice for fine aggregates.

The *bulk density*, or weight per unit volume, of coarse and fine aggregate is determined for an oven-dry sample using the method of Experiment 12, Volume I, but see Table 8, page 306, for details of container size and compaction of sample. From this, and knowing the specific gravity (oven-dried basis), the *percentage voids* may be calculated from:

$$\frac{\text{Voids}}{(\%)} = \frac{(\text{s.g.} \times 62\cdot4) - \text{bulk density (lb/ft}^3)}{\text{s.g.} \times 62\cdot4} \times 100$$

In metric terms this would be

$$\frac{\text{s.g.} - \text{bulk density (kg/dm}^3)}{\text{s.g.}} \times 100$$

The *percentage bulking* of a damp sample is calculated from:

$$\frac{\text{Bulking}}{(\%)} = \frac{\text{dry bulk density} \times (100+n)}{\text{uncompacted damp bulk density}} - 100$$

where n = percentage moisture content of the damp sample. Various methods are given for the determination of *moisture content*.

An alternative method (not a BS test) of finding the voids content of an aggregate, which does not require the specific gravity to be known, is given in Experiment 14, Volume I.

The BS test for *water absorption* is combined with that for specific gravity, and relates to the saturated surface-dry (s.s.d.) and oven-dry conditions:

$$\text{Water absorption} = \frac{\text{weight s.s.d.} - \text{weight oven-dry}}{\text{weight oven-dry}} \times 100$$

The procedure used to determine water absorption given in Experiment 23, Volume I (referred to there as the *absorbed moisture*), although different in sequence to that of BS 812 gives an equivalent result for fine or coarse aggregates.

11 *Organic impurities in fine aggregate*

This is a laboratory test involving the determination of the pH value of a prepared cement mortar containing the aggregate.

12 *Silt content*

BS 812 gives a *field settling test* as a guide to the percentage of silt, clay, and fine dust in natural sands. Laboratory methods given are the *decantation test*, which determines the amount of a sample to pass a BS 200 mesh sieve (75 microns), and a *sedimentation test*, which gives the amount of material less than 20 microns (0.02 mm) particle size.

EXPERIMENT 15 Field settling test for sand (silt test)

Note This test is not applicable to crushed stone sands or coarse aggregates

Apparatus Measuring cylinders (250 ml and 100 ml)—beaker (250 ml) and stirrer—physical balance (metric)—rule

Materials Common salt (sodium chloride)

Specimen Natural sand
1 Prepare a 1% solution of salt in water
2 Pour about 50 ml of the solution into the 250 ml measuring cylinder
3 Add sand up to the 100 ml mark
4 Add more salt solution until this reaches the 150 ml mark
5 Shake the mixture vigorously and leave it to settle in the measuring cylinder for 3 hours
6 Measure the thickness of the visible silt layer which has formed at the surface.
7 Measure the depth of sand, excluding the silt layer

Results Express the thickness of the silt layer as a percentage of the depth of the sand (see Example 11). Check the result obtained against the BS limit given in the footnote to Table 9, page 307.

Note Where a result is excessive a sample would normally be submitted for laboratory tests (for example, the standard decantation or sedimentation method of BS 812) before considering its approval for the proposed work.

Example 11 From the following results obtained in a field settling test on a natural sand, calculate the silt content.

Thickness of visible silt layer = 5 ml graduations
Depth of sand = 98 ml graduations
Silt content,

$$\text{per cent by volume} = \frac{5}{98} \times 100$$
$$= 5 \cdot 1\%$$

EXPERIMENT 16 Aggregate passing a BS 200 sieve (75 microns)

Introduction This is a laboratory method of test to establish the amount by weight of very fine material (as silt, loam, clay, etc.) present in a natural aggregate (coarse, fine or all-in). All-in aggregates may be separated into coarse and fine components which are both tested, and a combined result calculated on a pro-rata basis, according to the fractions retained and passing the relevant mesh size respectively. To avoid overloading sieves, testing in increments may also be necessary

Apparatus BS test sieves 200 (75 microns) and 14 (1.20 mm)—pan—physical balance—ventilated drying oven ($105 \pm 5°C$)

Specimen Sand or coarse aggregate, oven-dried then cooled, and of sufficient amount to provide a minimum *sample weight* as follows:

Nominal maximum aggregate size	$2\frac{1}{2}$–1 in (63·5– 25·4mm)	$\frac{3}{4}$–$\frac{1}{2}$ in (19·05– 12·7mm)	$\frac{3}{8}$–$\frac{1}{4}$ in (9·53– 6·35mm)	$\frac{3}{16}$ in or less (4·76 mm max.)
Minimum dry lb	13	2	1	0·5
weight of sample kg	6	1	0·5	0·2

Method

1 Weigh the dried sample (A)

2 Fit the two sieves together (larger mesh on top) and place them over a sink

3 Place the sample in the pan, cover it with water and agitate it vigorously, then pour the wash water through the sieves, avoiding as far as possible decantation of the coarser particles

4 Add further wash water and repeat the agitation and decantation procedure, continuing the process until the wash water runs clear
5 Return the material retained on the sieves (if any) to the washed sample
6 Dry the sample in an oven, cool it then re-weigh it (B)

Result Material finer than BS 200 (75 microns) sieve (per cent)

$$= \frac{A-B}{A} \times 100$$

The result obtained may be compared to the BS limits given in Table 9, page 307.

EXPERIMENT 17 Specific gravity and absorption of coarse aggregate larger than $\frac{3}{8}$ in (9·5 mm)

Note The method given here varies from that of BS 812 : 1967 (the BS test requires oven-drying to follow the remainder of the test, whereas it is more convenient to oven-dry at the commencement)

Apparatus Wire basket $\frac{1}{4}$ in (6 mm mesh) to hold at least 2 kg of the sample, and a watertight container in which to suspend the basket —tray—scoop—drying cloths—physical balance (3 kg × 0·5 g) with provision for weighing a sample suspended in water

Specimen $\frac{3}{4}-\frac{3}{8}$ in (19–9·5 mm) (sieved) coarse natural aggregate, washed, oven-dried at 105 ± 5°C, then cooled in an airtight container

Method
1 Weigh out not less than 2 kg of the dry sample (w)
2 Place the sample in the basket, submerge it in water, jolt it 25 times to displace air and leave it immersed for 24 hours
3 Repeat the jolting then weigh the sample, together with the basket, suspended in water (A)
4 Weigh the basket alone (after jolting), suspended in water (B)
5 Bring the aggregate to the saturated surface-dry condition and weigh it alone in air (W)

Results

Weight of sample, s.s.d., in water (A − B)   =—g (C)

Specific gravity on an oven-dried basis $= \dfrac{w}{W-C}$

$$\text{Specific gravity on a saturated surface-dry basis} = \frac{W}{W-C}$$

$$\text{Apparent specific gravity} = \frac{w}{w-C}$$

$$\text{Water absorption, per cent of dry weight} = \frac{W-w}{w} \times 100$$

where w = weight of oven-dried aggregate in air (in g)
W = weight of s.s.d. aggregate in air (in g)
C = weight of s.s.d. aggregate in water (in g)

EXPERIMENT 18 Specific gravity and absorption of aggregate smaller than $\frac{3}{8}$ in (9·5 mm)

Note The method given here varies from that of BS 812 : 1967 (for the reason stated in the note to Experiment 17)

Apparatus Pycnometer (see Fig 21)—(a vessel of 1 litre capacity with a metal conical screw top with a 5 mm diameter hole at its apex, giving a watertight connection)—warm air blower—tray—scoop—drying cloth—physical balance (3 kg × 0·5 g)

Specimen Natural sand (oven-dried, then cooled in an airtight container)

Method

[*a*] *Calibrating the pycnometer*

1 Remove the cap and fill the jar with water

2 Replace the cap tightly and top up with water until it overflows and runs free of air bubbles

3 Wipe the outside of the filled pycnometer, dry it, then weigh it (P_1)

[*b*] *Test*

4 Weigh out 500 g of dry sand (w)

5 Immerse the sample in water for 24 hours (agitate it occasionally to release air)

6 Bring the aggregate to the saturated surface-dry condition and weigh it in air (W)

7 Fill the pycnometer jar to one-third of its depth with water, add the sample (s.s.d.), replace the screw-cap and top up with water until it overflows and runs free of air bubbles. Air removal is assisted by rotating the pycnometer, on its side, then standing it upright

8 Wipe the outside of the filled pycnometer dry, then re-weigh it (P_2)

Results

$$\text{Specific gravity on an oven-dried basis} = \frac{w}{W - (P_2 - P_1)}$$

$$\text{Specific gravity on a saturated surface-dry basis} = \frac{W}{W - (P_2 - P_1)}$$

$$\text{Apparent specific gravity} = \frac{w}{w - (P_2 - P_1)}$$

$$\text{Water absorption, per cent of dry weight} = \frac{W - w}{w} \times 100$$

where w = weight of over-dried aggregate in air (in g)
W = weight of s.s.d. aggregate in air (in g)
P_1 = weight of pycnometer filled with water (in g)
P_2 = weight of pycnometer, sample and water to fill (in g)

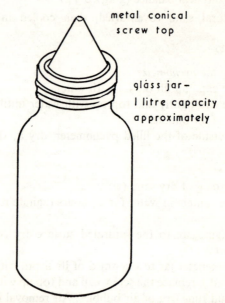

metal conical
screw top

glass jar –
I litre capacity
approximately

Fig 21 A pycnometer for specific gravity tests

EXPERIMENT 19 Aggregate crushing value (ACV) test

Apparatus (Fig 20) Standard hardened steel test cylinder [open-ended, 6 in/154 mm internal diameter, 5 in/130 mm height] with plunger and baseplate, to BS 812—steel tamping rod, $\frac{5}{8}$ in (16 mm) diameter, 24 in (600 mm) long with rounded end—cylindrical measure, $4\frac{1}{2}$ in (115 mm) diameter, 7 in (180 mm) height (internal)—BS sieves $\frac{1}{2}$ in (12·7 mm), $\frac{3}{8}$ in (9·53 mm) and BS 7 (2·40 mm), balance (3 kg × 1·0 g)—compression testing machine (capacity 50 tonf or 50 tonnef, or 500 kN)

Specimen Gravel or crushed stone aggregate, previously air-dried, sieved to obtain the $\frac{1}{2}$–$\frac{3}{8}$ in (12·7–9·5 mm) fraction. This must be oven-dried at 105°C for four hours then cooled to room temperature before testing.

Method

[a] *Sample measure*

1 Fill the cylindrical measure with aggregate, in 3 equal layers, each rodded 25 times and the final layer levelled flush with the rim
2 Weigh the measured aggregate (A)

[b] *Test procedure*

3 Fit the test cylinder onto its baseplate and add the measured sample in three equal parts, each rodded 25 times
4 Level the surface of the aggregate and fit the plunger
5 Place the assembled apparatus in the compression testing machine and load at a uniform rate to reach the maximum of 40 tonf (40·64 tonnef or 398·6 kN) in 10 minutes
6 After releasing the load, remove the whole of the test sample from the cylinder and sieve it to determine the amount which passes the BS 7 (2·40 mm) sieve (weight B)
7 Repeat the test on a duplicate sample with the same weight as the first sample

Result These are calculated from:

$$\frac{B}{A} \times 100\%$$

The two results are each calculated to the first decimal place and the mean value stated to the nearest whole number as the *aggregate crushing value* (ACV)

Note Where the ACV is 30% or higher, the result is held to be anomalous and the *10% fines* value should be determined and reported instead. The procedure is similar to that for the ACV test given here, except that a standard *distance of penetration* of the plunger (with uniform loading rate in 10 minutes) is specified in lieu of standard loading conditions. This distance must be provisionally about:

0·60 in (15 mm) for rounded or partially rounded aggregates
0·80 in (20 mm) for normal crushed aggregates
0·95 in (24 mm) for honeycombed aggregates

but subsequently adjusted to that giving between 7·5% and 12·5% passing the BS 7 (2·40 mm) sieve for the crushed sample. The result is given as the *load required for 10% fines*

$$= \frac{14 \text{ x}}{\text{y}+4}$$

where x = load in tonf,
and y = per cent fines (av. of 2) at x tonf load.

BS requirements for aggregates tested for 10% fines value are given in Table 10, page 307.

Summary

Aggregates are typically inert materials in granular form, mainly used together with a binder such as cement, plaster or bitumen.

They may be natural or crushed stone sands, natural gravels and crushed stone coarse aggregates. Alternatively there are numerous artificial and lightweight aggregates such as crushed brick, blast-furnace slag, clinker, pulverised fuel ash (p.f.a.), pumice, diatomaceous earths, exfoliated vermiculite, and the expanded forms of perlite, clay, shale and slate. Heavy aggregates include barytes, haematite and metallic iron or steel.

A grading test (or sieve analysis) is made by passing a representative sample of an aggregate through a series of test sieves of different mesh size, with the result expressed as the cumulative percentage passing each sieve mesh size.

The fineness modulus of an aggregate is an index of its grading, obtained by adding together the percentage weights of material retained on each of nine test sieves in a particular series and dividing by 100.

The surface area index of an aggregate is a single figure expressing the approximate relative total surface areas of different aggregates, obtained by allocating a surface area factor to each size of particle present and summing the products of the surface area factors and the percentage of material present of that size.

A fine aggregate can bulk when damp, due to the film thickness of water around its particles.

$$\frac{\text{Bulking}}{(\%)} = \frac{\text{damp (bulked) volume} - \text{dry volume}}{\text{dry volume}} \times 100$$

$$= \frac{\text{damp (bulked) volume} - \text{inundated volume}}{\text{inundated volume}} \times 100$$

$$= \frac{\text{dry bulk density} \times (100 + n)}{\text{damp bulk density}} - 100$$

where n = percentage moisture content

Tests given in BS 812 for aggregates include particle shape and size (flakiness index, elongation index and angularity number), specific gravity, water absorption, crushing and impact strength, abrasion and friction values, grading (sieve analysis) voids and density, bulking and moisture content, silt and organic contents.

5

MORTARS

Mortar may be defined as the bedding material used for bricks, blocks and masonry, in cases where the joints have significant thickness (3 mm or more). This serves to distinguish a mortar from the more fluid grout or slurry used to fill the irregularities of surface where units are butted against one another.

The three basic types of mortar are *cement mortar*, *cement-lime mortar* and *lime mortar*, where in each case sand is an added ingredient.

Properties of mortar
The desirable properties of a mortar are:
1 Good *workability* in the plastic state, but with sufficient early stiffening to prevent joints squeezing out
2 Good water *retentivity* (the mixing water not to 'bleed' out)
3 Adequate '*cohesion*' in the plastic state, to ease handling and reduce droppings
4 Adequate *adhesion* to the bedded units both in the plastic and hardened state. The term *bond* is used for the hardened state
5 Sufficient *strength* when hardened, for the work concerned
6 *Durability* sufficient with regard to the degree of exposure likely and any aggressive chemical agency
7 Tolerable *drying shrinkage* and *moisture movement* where this is relevant to the proposed work
In practice, an improvement in any one of these properties can readily be achieved by varying the mix proportions or constituents, but usually to the detriment of one or more of the other characteristics. For example, an increase in cement content results in a stronger mortar with greater drying shrinkage. The use of a finer

sand in a mix improves its cohesion, water retentivity and workability, but only for less strength and greater drying shrinkage. The
question of materials cost is also usually an important factor in these
considerations, resulting in a need to limit the cement content, which
is the costliest ingredient, and thus the strength, as far as possible.

Types of mortar

The strongest mortars are the *cement mortars*, but sand is incorporated and the ratio of sand to cement should not normally be less than
about three to one. Such mixes are confined to work which is below
the ground level damp-proof course, to very exposed work, such as
exterior free-standing walls and parapets, and to engineering construction with bricks of high strength; all are cases where the main
need is for a dense mortar of high strength and low permeability.
For other types of work, a strong mortar is both unnecessary and
undesirable. Unfortunately, the use of much weaker cement mortars,
which are mortars of lower cement content, is not practicable, since
any notable reduction of cement content leads to reduced *workability* (harshness) and less *cohesion*, and will produce a *porous*
joint with a tendency for *low frost resistance*. However, these de-
ficiencies can be rectified by the inclusion of a proportion of lime in
the mortar, and this explains the importance of *cement:lime:sand*
mortars. Another advantage of lime is that it increases the water
retentivity of mortar, which prevents the excessive withdrawal of
water, necessary for laying the units and for the hydration of the
cement, by units with high suction. By including lime the ratio of
sand to cement can be made twice that of a cement mortar. In fact,
the beneficial effects of lime may even be desirable when using the
1:3 cement:sand mortar referred to earlier, and this mix is often
'gauged' with up to one-quarter part of lime.

An alternative to the use of cement:lime:sand mortars is the use of
a hydraulic or, more usually, a semi-hydraulic lime with sand. This
type of mix is known as *lime mortar*, and one of its characteristics
is the continued strength development over long periods. This is
considered to be a useful property in mortars for tall chimneys.
However, for high early strength a cement-based mortar is preferable.

Lime mortars made with non-hydraulic lime can only harden very
slowly by carbonation and are unsuitable for mortars except in
special cases; for example, for internal work with very thin mortar

joints, or externally in sheltered conditions where the mortar is protected by a frost-resistant pointing.

Other alternatives to cement:lime:sand mortars are plasticised, or aerated, mortars, and the use of masonry cements (see Chapter 1). In the case of *plasticised mortar* lime is replaced as a workability aid by a proprietary plasticiser, added to the mix in small quantity in powder or liquid form, usually in the mixing water. These plasticisers are mainly wetting agents which act by reducing the surface tension of the mixing water, or air-entrainers.

Proportioning mortars

It is customary for the mix proportions of mortar to be stated as parts by volume. Examples of typical proportions for brickwork and block-work mortars to meet various requirements and conditions are given in Table 11 on page 308. From this you will see that the strength requirements for mortars depend on the weather conditions during the construction and the degree of exposure to which the finished work will be subjected. Table 12 on page 309 gives mix proportions for mortars for masonry. Since in most classes of masonry the mortar joints are relatively thin, the weather conditions are of less consequence in determining mix proportions, and the main factors are the porosity and strength of the stone. For example, the stronger mortars are applicable only for stone of high density and strength. For appearances, the sand used for masonry mortars is often crushed stone of the same variety as the units to be bedded, and it is also common for white or coloured cements to be used.

For units liable to undergo high drying shrinkage it is important to avoid the use of very strong mortars, unless permanently damp conditions are expected, in which case the drying shrinkage will be suppressed. This relates mainly to certain types of lightweight concrete blocks and some classes of sandlime bricks, for which mortars stronger than about 1:2:9 cement:lime:sand should not be used for internal work. If stronger mixes are necessary, units with low drying shrinkage should be specified. In any event, adequate curing of concrete blocks is essential before they are used. For natural curing in air one month is recommended. Such units should not be laid wet and any necessary initial adjustment of suction by damping should be kept to a minimum.

Mortar for tall chimneys should be limited to the weaker mixes to

meet the special conditions which apply (they more readily accom-
modate slight wind and thermal movements), and suitable mixes
are 1:2:9 cement:lime:sand, 1:3 hydraulic lime:sand or a comparable
mix with masonry cement; 1:6 would be typical. Where internal units
based on gypsum are to be bedded, a mortar with gypsum plaster as
binder may be used.

Mortar plasticisers. Masonry cements

Mortar plasticisers, when used as an alternative to lime, permit the
use of mortar mixes with aggregate:cement ratios comparable to
those of lime-gauged cement:sand mixes for the same purpose.

Masonry cements give adequate workability when used only with
sand although the amount of sand added is usually less than the
amount added for a lime-gauged cement mortar to serve the same
purpose.

In all cases where either plasticisers or masonry cements are used,
it is important to follow the manufacturer's instructions regarding
mix proportions, especially in relation to work in exposed positions.

Mixing mortars

Mortars can be mixed by hand or by mechanical methods. Uniform
dispersion of the materials is essential, with just the right amount of
water to achieve the required workability.

Any mix containing cement should normally be used up within two
hours of mixing, and the addition of further water after first mixing,
known as *retempering*, should not be permitted. However, *remixing*
(agitation without the addition of water) assists in maintaining
workability of the mortar within the permissible working period,
without detriment to the final strength of the mortar.

It is recommended practice to steep non-hydraulic hydrated lime
in water for 24 hours or longer before use to obtain the best workability. The lime should be added to the water, not the water to the
lime. In the case of lime putty prepared from quicklime this will, of
course, already have been done. The cement and sand are then mixed
together (dry) and blended with matured lime putty (with extra water
to give the required consistency) immediately before use. Alternatively, ready mixed lime:sand mortar (*coarse stuff*), consisting of a
mix of non-hydraulic lime and sand with water, may be delivered to

the works and kept available to be gauged with cement immediately before use, and water added to the required consistency. This must be distinguished from *ready mixed cement:lime:sand mortars*, which are fully mixed with water and consequently must normally be used within an hour or so of delivery. Experiments have been made to extend the normal working period up to 24 hours or more by incorporating a retarder to delay the setting of the cement, but such mixes are unlikely to satisfy a conventional specification. Coarse stuff is usually supplied at proportions 1:3 lime:sand, and is adjusted to the finished proportions as required. For example:

For a 1:$\frac{1}{4}$:3 (cement:lime:sand) mix from 1:3 coarse stuff—mix *one* volume of the coarse stuff with *three* volumes of sand (giving a 1:12 mix) then add *one* volume of cement to *three* volumes of the sanded coarse stuff. The lime does not increase the volume of sand but merely occupies space between the grains.

For a 1:1:6 (cement:lime:sand) mix from 1:3 coarse stuff—mix equal volumes of coarse stuff and sand (giving a 1:6 mix) then add *one* volume of cement to *six* of the sanded coarse stuff.

For a 1:2:9 (cement:lime:sand) mix from 1:3 coarse stuff—mix *two* volumes of coarse stuff with *one* of sand (giving a 1:4$\frac{1}{2}$ mix) then add *one* volume of cement to *nine* volumes of the sanded coarse stuff.

Note Where coarse stuff is prepared at the works it is usually mixed directly to the final proportions given above for the sanded coarse stuff (1:12, 1:6 or 1:4$\frac{1}{2}$). These are then gauged with cement as indicated.

Apart from ready mixed mortars, there are *premixed mortars* available which consist of the blended dry materials and only need water added before use.

Aggregates for mortar

The BS grading requirements for sands for mortar are given in Table 13 on page 309.

Cold weather precautions

Mortars are most vulnerable to frost attack during the setting period and the early stages of hardening, when free water present can freeze, causing expansion and consequent disruption of the work. The use of the richer, and therefore stronger, mixes recommended for winter conditions is mainly to reduce the likelihood of this occurring. If the units to be bedded are themselves kept dry and preferably warm, and the newly erected work is covered to keep it both dry and insulated against the cold, there will be still less chance of freezing. It is essential also that the sand used in the mortar shall be free of frost, ice or snow, and stockpiles should be properly covered. The use of a rapid hardening cement will also help by counteracting the retarding influence which low temperature has on the setting and hardening of mortar.

In extreme conditions it might be necessary to resort to the heating of materials before mixing, using the methods outlined for concreting (see Chapter 8).

The use of accelerators, such as calcium chloride or one of the proprietary 'frost-proofers', usually results in accelerating the setting and hardening of the mortar. The resulting increased rate at which the heat of hydration of the cement is evolved is of little consequence in view of the small volume of the mortar joints in relation to the bricks or blocks themselves. Also, calcium chloride and other salts can contribute to efflorescence and even dampness in the work if they are deliquescent (see Volume 1, Chapter 6).

EXPERIMENT 20 Crushing strength of mortars

Note This experiment assumes facilities for crushing tests on mortar cubes. An alternative test would be to prepare and test transverse specimens as shown in Fig 5, for modulus of rupture

Apparatus Mixing board—tray, gauging trowel—cube moulds (for example, 4 in or 100 mm size)—physical balance (metric)—brush—mould oil

Materials Ordinary Portland cement and dry building sand

Method

1 Check the moulds for correct assembly and security, then oil the inner surfaces, including baseplate

2 Prepare sufficient of each of the following mixes to cast, say, 3 cubes:

Mix	Proportions by weight	Materials for one cube (*4 in or 100 mm size*) Cement (*g*)	Sand (*g*)
A	1:3	650	1950
B	1:4	450	1800
C	1:5	350	1750

3 Mix the materials for each mortar together, first dry, then adding just sufficient water to produce an acceptable workability (mix C may be harsh). Note the amounts of water used

4 Make cubes from each mix and mark them for later reference[1]

5 Cover the cubes with plastics sheeting and leave them for 24 hours

6 Demould the cubes and store them in damp sand, or in a suitable humidity cabinet until tested at 7 or 28 days

7 Test the cubes with the side-cast faces in contact with the bearing plates of the testing machine

Results Tabulate these to show the water:cement ratio and average crushing strength of the three mixes

Note The results will show the effect on the strength of varying the cement content of mortars. The actual strengths will depend on the water:cement ratios used and the degree of compaction of the mortar in making the cubes.

The influence of sand grading on water:cement ratios is significant, and this could be the subject of a further experiment, using sands of fine and coarse gradings.

EXPERIMENT 2 1 The air content of plasticised mortar

Apparatus Cylindrical metal measure, approximate capacity 600 ml (e.g. 80 mm diameter × 120 mm height)—measuring cylinder (200 ml) —physical balance (to 0·5 g)—gauging trowel—mixing board—tray —drying cloth

Materials Natural sand (dry)—O.P. cement—mortar plasticiser (liquid)

Method

[a] Preparing a 1:5 (by weight) cement:sand mortar

1 Weigh out 200 g of cement and 1000 g of sand and mix them together

2 To 50 ml of water add plasticiser appropriate to the 200 g of cement taken (1 ml of plasticiser for the typical product requiring 0·5 Imperial pint per cwt of cement or 0·3 l per 50 kg of cement)

3 Mix the prepared solution together with the dry materials, then add

---

[1] There is no standard method of filling cube moulds with mortar, but filling in 3 layers and tapping the outside 20 times at each layer with a light rubber mallet, is one method. Alternatively, the cubes may be vibrated or manually tamped.

a further measured quantity of plain water, sufficient to make the mortar workable with continued mixing

[b] Air content
1 Weigh the cylindrical measure, empty
2 Fill the mortar into the measure in about 10 equal layers, each time dropping it on the bench several times to eliminate air pockets
3 Strike the mortar off level at the top of the measure
4 Weigh the filled measure to find the weight of mortar in it
5 Empty the measure, rinse it, then find its capacity by filling it with water

Result Calculate this from:

$$\text{Air content (\%)} = \frac{G - G_a}{G} \times 100$$

Where G is the mean absolute density (in g/ml) or specific gravity of the mortar, *calculated* from the known mix proportions and specific gravities,[1] and G_a its *measured* density in g/ml, or apparent specific gravity, from the experiment.

Note The air content obtained can vary with the amount of mixing, sand grading and temperature.

Example 12 A small mortar mix for test purposes was made using 200 g of cement (s.g. 3·10), 1000 g of sand (s.g. 2·60) and 160 ml of water (1 ml of plasticiser included). The amount of the mortar to fill a container of 600 ml capacity was 1200 g. Calculate the air content of the mortar.

Mix proportions by weight = 200 : 1000 : 160
$$= 1 : 5 \cdot 0 : 0 \cdot 8$$

Parts by weight $\qquad = 1 + 5 \cdot 0 + 0 \cdot 8 = 6 \cdot 8$

Parts by volume
(dividing by s.g.) $\qquad = \dfrac{1}{3 \cdot 10} + \dfrac{5 \cdot 0}{2 \cdot 6} + 0 \cdot 8 = 3 \cdot 05$

Mean s.g. (G) $\qquad = \dfrac{6 \cdot 8}{3 \cdot 05} = 2.23$

Measured density (G_a) $\quad = \dfrac{1200 \text{ g}}{600 \text{ ml}} = 2 \cdot 0 \text{ g/ml}$

[1] Take specific gravities of cement 3·10 and sand 2·60 unless these are known. For accurate results they should be determined by tests.

$$\text{Air content} = \frac{G - G_a}{G} \times 100$$

$$= \frac{2 \cdot 23 - 2 \cdot 0}{2 \cdot 23} \times 100$$

$$= \frac{23}{2 \cdot 23} = 10 \cdot 3\%$$

Summary

Desirable properties of a mortar are good workability, water retentivity, 'cohesion' and adhesion (bond), adequate strength and durability and tolerable drying shrinkage and moisture movement.

Cement: sand mortars are very strong mixes which should only be used with very dense units or in permanently damp conditions (owing to high drying shrinkage).

The early strength of cement:lime:sand mortars depends mainly on their cement:sand ratio, which should therefore be related to the strength and density of the units to be bedded and to the degree of exposure of the work. The lime content provides workability and water retentivity, and improves frost resistance.

Lime mortars, containing semi-hydraulic lime and sand, gain strength more slowly than cement-based mixes.

Other factors which affect the strength are the water content of mixes, which should be kept as low as possible, and the sand grading. Use less sand with the finer varieties and with badly graded sands.

Plasticisers of proprietary makes may be used in place of lime, provided the strength requirements are met by using a suitable cement:sand ratio.

Masonry cements provide another alternative to the use of lime. They are an intimate mixture of Portland cement and finely ground chalk or siliceous material, sometimes with an added plasticiser or air-entrainer. Since the interground chalk or silica replaces part of the cement, a corresponding reduction in the amount of sand added is usual. Always follow the manufacturer's recommendations.

Any mix containing cement should normally be used up within two hours of first adding the water. Retempering should not normally be permitted.

Re-mixing a mortar restores workability without loss of strength.

Non-hydraulic lime is recommended to be steeped in water for at least some hours before use to obtain the best workability.

Ready mixed lime:sand mortar consisting of non-hydraulic lime and sand mixed with water is often kept available at the works to be gauged with cement as and when required.

Premixed mortars are blends of the dry materials of a mortar, requiring only the addition of water before use.

Cold weather precautions include the use of the stronger mixes recommended for a given type of work, with dry, and preferably warm, bricks or blocks and frost-free sand. Rapid hardening cement is useful. In extreme conditions, materials should be heated and mixed (water and sand). The new work should be covered to keep it dry and insulated against the cold.

Air content of mortars is given by:

$$\frac{G - G_a}{G} \times 100\%$$

where G = calculated mean absolute density (g/ml) or s.g.

G_a = measured density (g/ml) or apparent s.g.

6

EXTERNAL RENDERING AND FLOOR SCREEDS

RENDERINGS

The term *rendering* is correctly used in describing a surface application of a cement:sand or similar mix to an *external* wall or other surface, either to give it a good appearance or to make it weatherproof, or both. You must not confuse this with the term *rendering coat*, which is sometimes used to refer to the *first coat* applied in plastering on internal walls.

A rendering treatment may comprise one, two or even three coats. The first is termed the *first* or *render* coat, and the last is termed the *final*, or *finishing*, coat. The name *floating* coat is given to the undercoat immediately preceding the final coat. Variety in the finished appearance of a rendering may be achieved by incorporating coarse or special aggregate, or by texturing treatments.

Avoidance of crazing

It is now recognised that the use of strong, dense renderings does not produce good weathering qualities, as might at first be expected, since these are susceptible to shrinkage cracks, which allow water to penetrate and become trapped behind the rendering. This does not apply when a proportion of coarse aggregate is used, for example, in finishes such as *roughcast* or *pebble-dash*, which are consequently favoured where a high degree of impermeability is desired. However, these two treatments can only be applied on strong, rigid backgrounds, otherwise loss of bond is likely.

Very smooth, highly trowelled rendered surfaces are also to be avoided, since they are liable to *craze* (develop a network of fine hair cracks at the surface). Excessive trowelling results in the accumulation of a 'skin' of rich cement mortar of high water content at the surface of the work, known as *laitance*. The laitance has a high drying shrinkage relative to the remainder of the rendering, so that crazing occurs.

Workability

A proportion of high calcium lime is commonly incorporated in the mix in order to limit the cement content and achieve the desired workability and durability. The alternative use of proprietary plasticisers or masonry cements has not yet found general approval for external use in renderings. Hydraulic limes may be used instead of cement and high calcium lime mixtures, where they can give equivalent strength.

Background suction

For the wet mix to adhere to the background there must be a limited amount of suction; high suction is undesirable since the rendering must retain sufficient water to allow hydration of the cement. The reluctance of a mix to part with its water, its *water-retentivity*, is also important, and this varies for different mixes. The retentivity is generally less for leaner mixes, and should be increased by adding lime. In some cases the background suction may need to be reduced by a preliminary damping of the surface, but no visible water-film or droplets should be present at the surface when the rendering is applied, otherwise bond (adhesion of the hardened rendering) will be reduced.

Improving bond

The achievement of good bond between the hardened rendering and the background is facilitated by a rough-textured surface, rather than a smooth, dense, surface. In some cases a specially grooved (keyed) surface will be available to give a *mechanical key*, as distinct from the *specific adhesion* obtained without key. For smooth, dense backgrounds, one or more of the following methods may be adopted to promote good bond characteristics:

[a] the raking out of mortar joints to about ½ in (or 1 cm) depth
[b] hacking the surface to provide mechanical key
[c] the application of a preliminary *spatterdash* treatment. This is a thin coat of a cement sand slurry, of proportions 1 : 2 approximately,

thrown on wet to form an uneven layer $\frac{1}{8}$–$\frac{1}{4}$ in (3–6 mm) thick. The spatterdash will also assist weather-proofing. The background must be damped before the application, and a further damping should follow some hours after the application to ensure hydration. A coarsely graded, sharp (angular-grained) sand is needed for this. [d] The preliminary application of a proprietary bonding agent, many of which are based on PVA (polyvinyl acetate) solutions. Alternatively, the bonding agent may be added to the mixing water of the render coat. Similarly, it is sometimes mixed in with a spatter-dash.

A method sometimes used on very difficult surfaces and in renovating old work is to apply the rendering onto metal lathing or expanded metal ($\frac{3}{8}$ in/10 mm or $\frac{1}{4}$ in/6 mm mesh) fixed to battens plugged to walls, or onto small-mesh galvanised wire netting fixed against the wall face.

Undercoat

An undercoat is normally between $\frac{3}{8}$ in (10 mm) and $\frac{5}{8}$ in (15 mm) thick. It is 'combed' or 'scratched' before hardening to provide a key for the subsequent coat, except where one of the thin machine-applied finishes is to be used. The combing should leave a wavy pattern of scratch marks in the horizontal direction about $\frac{3}{4}$ in (20 mm) apart, penetrating to a depth not greater than half the thickness of the coat.

If the first undercoat does not provide a sufficiently good surface to receive the final coat directly a second undercoat ($\frac{3}{8}$–$\frac{1}{2}$ in/10–13 mm thick) may be applied. This is often the case on very irregular backgrounds. A second undercoat is also needed on metal lathing or expanded metal due to penetration of the first coat, and where the final coat is to be relatively thin (say $\frac{1}{8}$ in/3 mm) so as to provide adequate total cover for weather resistance.

Mix proportions for the undercoat should be chosen to suit the suction and rigidity of the background. The stronger mixes are suitable only for strong, rigid backgrounds. Mix proportions commonly used are 1:3 cement:sand (with up to $\frac{1}{4}$ part lime), 2:1:8–9 cement:lime:sand, 1:1:5–6 cement:lime:sand, or 1:2:8–9 cement:lime:sand (or 1:2–3 hydraulic lime:sand). The 1:2:8–9 mix, or equivalent, is not generally suitable for work to withstand severe exposure. All mix proportions given are by volume and those for lime refer to lime putty (lime pre-soaked at least overnight in water). Where the dry hydrate is used instead of putty, the proportion of lime can be increased by half to give the same workability. The makers of mortar-plasticisers give their own recommendations for rendering mixes, but, as with mortars, these plasticisers replace the lime and the proportions of cement and sand are comparable, but usually with reduced sand content.

Final coat

This will not normally be more than half the thickness of the preceding coat, and should not be stronger, except where coarse aggregate is included. Examples of different finishes follow, and in each case the lime can be replaced by another plasticiser, or a masonry cement can be used, in each case with the manufacturer's recommended mix proportions. Coloured cement can also be used.

Adequate drying time should be allowed between successive coats so that drying shrinkage is not transmitted through to the final coat.

1 *Floated finish* A final coat of about $\frac{1}{4}$–$\frac{3}{8}$ in (6–10 mm) thickness is applied, and brought to a level surface with a wood or felt-faced float. Excessive working should be avoided to reduce the risk of crazing and the use of a coarse sand also helps. Mixes mainly used are 1:1:5–6 cement:lime:sand for fully exposed surfaces, to 1:2:8–9 cement:lime:sand on sheltered surfaces. Stronger mixes can be used where high abrasion resistance is needed, provided the background is strong and dense and a strong undercoat is used.

2 *Textured or combed finish* Various patterned treatments are obtained by working the freshly applied final coat with different tools.

3 *Float textured finish* After applying the floated finish it is left to stiffen (about 2 hours depending on atmospheric conditions). The cement skin is then removed by pressing against the surface a felt-covered float, damped with water. The float must be frequently rinsed in water to remove the accumulation of 'laitance'.

4 *Scraped finish* This is a variation on the floated finish, obtained by scraping the surface between three to sixteen hours after application, with a metal scraper, saw-blade or float faced with expanded

metal. The variation in time depends upon the mix and the weather. The object is to remove the surface layer of cement skin to provide a granular surface texture with less tendency to craze than would otherwise be the case.

5 *Brushed finish* This is another process applied to a floated surface and it produces an effect similar to a scraped finish. A wire brush may be used after about sixteen hours, or alternatively a stiff-haired brush may be used with water at an earlier stage, before hardening has occurred.

6 *Pebble-dash* (*dry dash*) This finish is only suitable for strong rigid backgrounds. The *first coat* is usually a 1:3 cement:sand mix (with $\frac{1}{4}$ part lime), or 2:1:8–9 cement:lime:sand. Alternatively, lime may be omitted and a water-repellent cement or a waterproofing additive used. The *second* coat should be applied about $\frac{3}{8}$ in (10 mm) thick, of proportions 2:1:8–9 cement:lime:sand. Washed shingle or crushed stone, size $\frac{1}{2}$–$\frac{1}{4}$ in (13–6 mm) or $\frac{3}{8}$–$\frac{3}{16}$ in (10–5 mm), is thrown on (wet, after draining), and is sometimes lightly tapped with a float, so as to be left protruding from the rendering.

7 *Roughcast* (*wet dash*) The *first coat* should be of 1:3 cement:sand (with $\frac{1}{4}$ part lime), 2:1:8–9 or 1:1:5–6 cement:lime:sand, depending on the strength and density of the background. If a second coat is necessary before the final coat, either the next weaker of the three mixes is appropriate or the same mix if preceded by the 1:1:5–6. The final coat is a wet plastic mix of, for example, cement:lime:sand 2:1:3, with $1\frac{1}{2}$ to 2 parts of added crushed stone or shingle ($\frac{1}{2}$–$\frac{1}{4}$ in 13–6 mm). All constituents are thoroughly mixed and water is added to produce a plastic consistency. The roughcast is applied by being thrown against the wall by laying-on trowel or scoop, or by mechanical methods.

8 *Tyrolean* The work is usually done in three coats. The *undercoats* are of proportions 1:1:5–6 to 1:2:8–9 cement:lime:sand, according to the degree of exposure of the work. The *final coat* is of similar mix proportions but with specially graded sand of selected quality and colour, and usually a coloured cement. Alternatively the finishing material may be obtained premixed, ready for use when water is added. The finishing mix is applied by flicking or spraying by hand or with powered machines, often to a thickness of about $\frac{1}{8}$–$\frac{3}{16}$ in (3–5 mm).

Precautions in applying renderings

Rendering should be allowed to dry out naturally with the background, and should be protected from extremes of weather, direct sunlight or artificial heat. This will prevent undue differential movement of rendering and background from drying shrinkage and thermal variations, which cause cracking and loss of adhesion. The interval between the application of successive coats should preferably be at least two days in warm, dry weather, extending to a week or more in cold, wet weather. This allows the greater part of the initial drying shrinkage to take place without affecting the subsequent coat, which should be of a weaker mix.

Renderings to assist waterproofing

Various products which can be added to rendering mixes to impart water-repellent properties, or to reduce permeability, are available. Such additives may act variously by filling the pores of a rendering, by reacting chemically with the compounds present to seal the pores, or by effectively reducing the adhesion of the hardened material to water. These additives must be thoroughly dispersed throughout the wet mix to achieve satisfactory results. As an alternative to using these additives a water-repellent cement may be used.

Surface treatments may be applied to existing rendered surfaces to improve their weatherproofing characteristics, but the life of these treatments is questionable. Silicone solutions are said to give satisfactory results for a number of years after which a further application would be needed.

These renderings are sometimes used to reduce suction so as to facilitate subsequent treatments, or as a means of avoiding uneven suction. A typical example is where a coloured finish is to be applied in which case an even suction is necessary to avoid patchiness. Another example is in the rendering of no-fines concrete, which, due to its honeycombed structure, provides excellent mechanical key, but offers negligible suction to promote specific adhesion. Because of the open texture of the no-fines, a first coat tends to have a high and irregular suction, making it difficult to apply a second coat satisfactorily. Adjustment of the suction of the first coat by damping might prove to be adequate treatment, but a surer alternative is to use a water-repellent cement in the first coat.

D

Aggregate grading

The grading requirements of BS 1199 for sands for external rendering are given in Table 14, page 310. These grading limits are specially for undercoats and final coats, but are not used with roughcast and some tooled or textured finishes. For roughcast the grading of the fine aggregate must suit the grading of the coarse aggregate used, and also produce an acceptable, finished texture. Tooled finishes may require a finer, or a coarser, aggregate, depending on finished texture desired and the requirements of the mix for the particular tooling operation.

FLOOR SCREEDS

Floor screed is a term applied to the relatively thin layer of mortar laid on top of a floor slab to bring it to a required *level* (height) and *contour*, either flat (horizontal or inclined) or curved, to receive the floor finish. Floor finishes include tiles and sheeting of clay, cork, rubber, plastics, wood block and linoleum.

A screed should not normally be left to form the wearing surface of a floor, since it is unlikely to offer sufficient resistance to wear by abrasion. Appropriate materials in these circumstances might be granolithic concrete, terrazzo, floor quarries, metal-clad flags, a suitable grade of mastic asphalt or latex-cement composition, suitably hard brick, or a floor topping based on a synthetic resin with a suitable mineral aggregate.

Thickness of floor screed

The thickness of the floor screed may be governed by the finished floor level, but it should preferably relate to the condition and age of the receiving surface. Where exceptionally good bonding of the screed to a concrete base is obtainable, the minimum thickness required may be as little as $\frac{1}{2}$–1 in (13–25 mm). but this will normally be possible only when screeding onto a fairly level 'green' base which has not had time to harden substantially. For older, or irregular, surfaces, which are clean and offer good bond, a thickness of about $1\frac{1}{2}$ in (40 mm) should be adequate. For old surfaces offering poor bond, the thickness should be increased to 2 in (50 mm) and the surface hacked to give mechanical key. Adhesion may also be improved by the use of a bonding agent, for example a PVA emulsion, applied directly to the base surface or added to the mixing water. Where the base is weak or likely to compact, the floor screed in effect assumes a structural rôle and should be $2\frac{1}{2}$–3 in (65–75 mm) thick, possibly with a light mesh reinforcement.

Any screed isolated by being laid over a damp-proof membrane or insulating quilt should be $2\frac{1}{2}$–3 in (65–75 mm) thick.

Where pipes or cables are to be embedded in the screed the thickness of the screed should allow for adequate cover.

Bay sizes

If a large area is to be screeded, bays may be laid draughtboard fashion in areas normally not exceeding about 120 ft² (11 m²), but up to twice this area if laid on 'green' concrete. These bays should be left for at least 24 hours before screeding intermediate bays, which may normally be butt-joined without special provision for thermal expansion.

Mix proportions

The usual mix proportions are from 1:3 to 1:4 cement:sand by weight and a particularly dry mix is necessary. It should appear *just* moist. This minimum water content avoids excessive shrinkage and low strength, as well as producing a mix suitable for trowelling. Where steel pins are to be driven into the floor screed for fixing purposes, a weak mix of proportions 1:5 may be used. For screeds *exceeding* about $1\frac{1}{2}$ in (40 mm) thick it is preferable to include a proportion of coarse aggregate, of $\frac{3}{8}$ in (10 mm) maximum size, with mix proportions $1:1\frac{1}{2}:3$ by weight.

Lightweight screeds are sometimes used, either aerating the mix with special additives, or using a lightweight aggregate.

Aggregates

The use of sands which are very fine or badly graded should be avoided, since this gives a weak screed with poor adhesion. BS 1199 gives general requirements for natural and crushed stone sands for floor screeds, including grading (see Table 14, page 310). *Lightweight aggregates* are sometimes used to produce lightweight screeds for better thermal insulation or fire resistance, nailability, or, with some types, to obtain resilience or better adhesion.

Laying and finishing

The application of screeds follows the general procedure for rendering, except that screeds are normally applied in one layer.

The *surface preparation* for screeding is very important. For direct application to *concrete* all loose material, including dust, must be removed and the surface hacked for key if necessary. The sub-base should be damped down to the saturated *surface-dry* condition before the screed is laid.

Where the concrete base is fully hardened, a thin coating of cement grout is sometimes brushed onto the surface just before the screeding. This is said to improve the bond. The grout is a mixture of cement and water to the consistency of thin cream. But a bonding agent can be used instead of the cement and water mixture. If the screed is to be laid on bitumen, or similar membrane, sharp sand should first be scattered thinly over the surface and lightly pressed in to allow adhesion to be achieved.

The *stiffness* of the mix should be such that *tamping* or *ramming* is required, and the surface may be left with the tamped finish or *floated* to suit the requirements of the proposed floor finish. The interval between laying the screed and first floating the surface varies, but it will normally have stiffened sufficiently within three to five hours of laying. One or more further floatings may follow within the next few hours. A steel float is normally applied with heavy pressure. The surface, however, must not be overworked as this will cause *laitance*, a rich layer of cement and fine sand drawn to the surface, which is liable to craze and flake or powder off. Where the proposed floor finish requires a granular rather than a smooth-texture surface, trowelling may be done with a wood float.

The laying of bays is assisted by the use of temporary side rails, or edge strips of levelled screed.

Curing

Adequate curing is essential to prevent cracking and curling at the edges. The work should be covered with impervious sheeting or sprayed with a membrane curing compound (for example, PVA composition), and left for seven days or longer.

Since a crack will usually form along a line over a joint in the base it is best to continue such joints through to the surface. This type of cracking is called 'sympathetic' cracking.

Applying floor finishes

Finishes such as clay tiles may be bedded in a thin mortar layer of 1:3 cement:sand laid directly onto the screeded base (*bonded*), or alternatively the bedding mortar may be separated from the floor screed by a sheet of building paper (*unbonded*). The unbonded method has the advantage that disruption due to differential movement of the screed and mortar bed is unlikely.

Finishes such as rubber flooring or thermoplastic tiles are usually fixed by an *adhesive* onto a float-finished floor screed. In such cases it is important to allow for adequate drying of the screed. Otherwise the finish will lift and be ruined.

Summary

Renderings

External cement-based renderings are applied in one or more coats, depending on the nature of the surface and type of finish required.

Adequate intervals should be allowed between successive coats for natural drying to avoid troubles from shrinkage. The length of the intervals depends upon the weather conditions.

The use of strong, dense renderings should be avoided except in special circumstances, such as 'waterproofing' and below ground level and for special mixes like roughcast and pebble-dash. These mixes should only be applied on dense, rigid backgrounds.

The strength of the mix should be related to that of the background, and to that of previous coats.

High calcium lime is normally added to cement-based mixes to impart workability, water retentivity and frost resistance.

Surfaces to be rendered should offer adequate suction for good adhesion to be obtained, or good mechanical key. Alternatively a spatterdash treatment or bonding agent can be used.

In addition to plain, floated finishes there are many special finishes including textured, brushed, combed or scraped finishes, pebble-dash (dry dash), roughcast (wet dash) and Tyrolean. Coloured and water-repellent cements are often used, and aggregates may be specially selected for colour and grading.

Floor screeds

A floor screed is a layer of mortar used to bring a floor to the level required to receive the floor finish, for example, tiles, wood block or sheeting.

To avoid a screed curling at the edges or becoming detached from a concrete base, the minimum thickness of a bonded screed should relate to the age and condition of the concrete base ($\frac{1}{2}$–1 in/13–25 mm on green concrete; $1\frac{1}{2}$ in/40 mm on matured concrete offering good bond). Adhesion is improved by the use of a bonding agent, or a coating of cement grout.

The screed is usually laid in draughtboard fashion with plain butt-joints and the surface should be damped beforehand.

On a weak base or one offering poor key, the floor screed should be unbonded (laid on building paper), and $2\frac{1}{2}$–3 in or 65–75 mm minimum thickness. At this thickness a larger maximum size of aggregate ($\frac{3}{8}$ in/10 mm) is preferable.

The screed should be floated or trowelled a few hours after laying, depending on weather conditions, and again at intervals as necessary.

Thorough curing of floor screeds is essential.

7

INTERNAL PLASTERING

The object of plastering is to provide a finished surface for back-grounds. Internal plastering is usually classified as plastering which is based principally on one of two binders; *Portland cement* or *gypsum plaster*. Hydraulic limes are alternatives to Portland cement. In each case *sand* and/or *lime* may be added. However, in adopting these simple classifications you should remember that in particular cases the amount of lime in the mix may exceed the amount of the binder. For example, a final coat might consist of non-hydraulic lime gauged with a small proportion of gypsum plaster. Also, there are some special *thin-wall* plasters in use, which consist of finely ground mineral substance with an organic binder such as polyvinyl acetate (PVA).

Materials used

The main references in this book to the individual materials used in plastering are given in Chapter 1 (cements), Chapter 2 (gypsum plasters), Chapter 3 (limes) and Chapter 4 (aggregates).

The BS grading requirements for sands for internal plastering are summarised in Table 15, page 310.

Plaster mixes

An outline of mix proportions for plastering is given in Table 16, page 311. Information in greater detail is given in BSCP 211 'Internal Plastering', or is obtainable from manufacturers.

Where sand is added to the mix the proportion used can be greater for application to absorbent, open-textured surfaces with good key than for smooth, dense surfaces. For example, less sand is normal for dense concrete and plasterboard surfaces. A reduction in sand content should be made when using very fine sands (for example, type 2 sands to BS 1198) and for badly graded sands (for example, predominantly single-sized).

Gypsum undercoat plasters are available with a hair of fibre content to give improved 'cohesion' and easier application (with reduced droppings) to ceilings and to surfaces with low or variable suction (including metal lathing). Metal lathing which is to be covered by gypsum plasters may be galvanised or lacquered to reduce the risk of corrosion. Alternatively, a special grade of gypsum plaster containing a corrosion inhibitor may be specified, or a lime addition about 10% by volume can be made to produce the same effect.

Application of plaster

Of fundamental importance in plastering work is the fact that cement-based plasters undergo prolonged, gradual, *drying shrinkage*, whereas gypsum plasters only undergo an *expansion* during a relatively short *hydration* period. Before applying successive coats to a cement-based undercoat, adequate drying time is therefore essential, otherwise cracking or bond failure may occur. No such delay is necessary between successive coats of gypsum plaster, and it is quite common for final coats to be applied on the same day as the undercoat. In fact this is desirable except in damp weather or with low porosity backgrounds, where an interval of a day may be required to allow the undercoat to dry out to the right degree of porosity to suit the next application.

Number and thickness of plaster coats

The number of plaster coats applied to a surface may be one, two or three, depending mainly on the regularity of the background. Consequently, the total finished thickness of plastering varies; undercoats are from about ½–¼ in (13–6 mm) thick and final coats about ⅛ in (3 mm) thick.

One-coat work can only be successful on level backgrounds such as plasterboard, some wallboards or fair-faced concrete. All coats except the final coat may be referred to as *undercoats*, and a first, or single, undercoat is also known as the *backing coat*.

Choice of binder

Cement-based mixes can provide extremely hard, impact-resisting surfaces. They can withstand permanently damp conditions unlike gypsum plasters, which are slightly soluble in water.

In general, cement-based plaster is used mainly for undercoats, whereas gypsum plasters are widely used both as undercoats and final coats.

Cement-based mixes, because of their drying shrinkage, are not suitable for application to non-rigid backgrounds, gypsum plasterboards, or materials liable to high moisture movement.

Cement and gypsum plaster must never be mixed together, owing to the adverse chemical interaction which could occur. However, it is quite acceptable, and usual, for a plaster coat based on gypsum to follow one based on cement. Always allow a suitable interval for drying, but never cover a plaster coat based on cement with a plaster coat based on gypsum because of the drying shrinkage of cement mixes.

Working times of gypsum plasters

The principal differences between gypsum plasters for undercoats and gypsum plasters for final coats are the fineness, setting times and finished hardness, which are adjusted in manufacture to meet requirements for application and finish.

Gypsum plaster setting times are stated by the manufacturers; they vary, but are commonly between one and two hours. Anhydrous plasters (Class C and D) often have a longer setting time than retarded hemihydrate (Class B) types, and this allows extra working time to bring them to a smooth, flat surface. Working times will vary also according to the mix proportions, especially lime or sand content, and the amount of mixing.

Class C (anhydrous) plasters usually undergo an early initial stiffening, followed by a slow and continuous set, and it is quite in order to *re-temper* the mix, that is, to adjust its consistency by mixing in more water, immediately after the 'false set'. No re-tempering should be necessary with Class B (hemihydrate) types.

Hardness of gypsum plaster

The finished hardness of plasters based on gypsum depends on whether the hemihydrate or the anhydrous type is used, and on the amount of sand or lime addition, as well as on the water content of the mix. In general, the hemihydrates give less hardness than the anhydrous varieties, and in all cases the hardness decreases as the lime content and mixing water is increased. Anhydrous types are normally used for final coats only. Therefore, the softest, most absorbent, finishes are the ones with a high lime content. The very hardest gypsum plasters (for example, Keene's Class D) are used mainly neat, as final coats at arrises and reveals (for example, around door and window openings) or at other points prone to mechanical damage.

In general, strong, hard finishes should not be applied on softer undercoats or non-rigid backgrounds, and in practice this limits the use of mixes of highest lime content to final coats.

Bond of gypsum plaster

Proprietory gypsum *bonding plasters* are available for application to concrete and similar smooth, dense surfaces. These may simply be 'low-setting expansion' types, or may incorporate a bonding agent. Alternatively, a bonding agent, such as a PVA (polyvinyl acetate) aqueous emulsion can be applied to the surface before plastering, or in the mixing water.

Premixed gypsum plasters

These are ready-blended mixes of the dry materials, and are ready for application as soon as they have been mixed with water. They include the *vermiculited* and *perlited* lightweight plasters, which offer greater resilience than normal density plasters, and so are useful on backgrounds liable to undergo slight movements.

Decoration of plastered surfaces

Only permeable decorative finishes unaffected by alkali are suitable for early application to cement-based plaster. Retarded hemihydrate gypsum plasters are not limited in this way, provided no lime is used in the final coat.

Summary

Mixes used for internal plastering consist usually of a fine aggregate and a binder, possibly with the addition of other materials as workability aids or to influence properties such as 'cohesion', water retentivity, porosity and surface texture or adhesion.

Materials used include gypsum plasters, cements, limes and organic binders, natural or crushed stone sands and special aggregates, such as exfoliated vermiculite or expanded perlite.

The number and thickness of plaster coats must take account of the nature and regularity of the background and the standard of finish required, and may be from one to three coats.

Cement-based mixes with lime provide strong, abrasion resistant surfaces and are durable in damp conditions, provided they are not exposed to chemical attack. They must be allowed adequate drying time before successive coats are applied because of their drying shrinkage. Early decoration is limited to permeable paint treatments not liable to alkali attack. They cannot be applied onto a gypsum plaster undercoat and are unsuitable for backgrounds which are non-rigid or undergo high moisture movement.

Gypsum-based mixes undergo a slight setting expansion. The amount varies with the mix proportions, but there is no delay between successive coats which are usually applied on the same day. Early decoration is possible but remember that premature drying out can result in failures due to delayed expansion.

Anhydrous plaster mixes which show a 'double set' can be re-tempered after the initial stiffening, but no retempering should be permitted for hemihydrate plasters.

Anhydrous plasters can give a harder finish than hemihydrate types and are used mainly for final coats.

Haired or fibred gypsum plasters are used for backing coats on backgrounds which are non-rigid or have low or variable suction.

Plasters of low setting expansion quality, or special 'bonding' plasters are used for smooth, dense surfaces without key; alternatively, a bonding agent may be used.

Premixed lightweight plasters are easily handled and applied and are useful for their special properties and resilience.

8

CONCRETE

Concrete is an artificial material, similar in appearance and properties to some natural limestone rock. It is formed by binding together particles of natural or artificial stone, brick or other aggregate with cement. One of the important properties of concrete is the ease with which it can be moulded to any shape before hardening occurs.

Normal concrete is distinguished from mortar by the fact that it contains coarse aggregate, in addition to fine aggregate and cement. However, there are special types of concrete which dispense with either the fine, or the coarse, aggregate. These will be dealt with under the heading of 'lightweight concrete'. Our first concern, however, will be the traditional type of concrete, made with normal density fine and coarse aggregates, cement and water. It is classified as *dense concrete*.

DENSE CONCRETE

Water content and voids

The presence of the coarse aggregate in concrete results in a hardened material which is far stronger than can be obtained using a mortar mix of the same cement content. This effect is achieved indirectly because less water is needed to produce a workable mix when coarse aggregate is added. This fact is fundamental in the study of concrete and will be explained further in the next few pages.

The setting and hardening of concrete is due to the hydration of the cement binder. This is a crystallisation process, and the aggregate can normally be considered as an inert constituent. The amount of water necessary for the complete hydration of 1 kg of ordinary Portland cement is roughly 0·25 kg but if these amounts of cement and water are mixed together, the resulting cement paste is barely workable. Evidently, if any aggregate were added to this mix, additional water would be needed to make it workable. This extra water, above that needed for hydration, acts as an extender of the cement paste, allowing it to coat the surface of all particles of aggregate and thereby to function as a lubricant, providing the *workability* necessary for handling and placing the mix. Unfortunately, the effect of this 'excess' water for workability is to form voids, or *pores*, in the hardened mortar or concrete, as the material dries out. These pores (*water-voids*) cause a reduction in density, strength and durability of concrete. It follows that the water content of mixes should normally be kept as low as possible, consistent with providing adequate workability to allow placing and full compaction. *Compaction* entails the removal of air from concrete by vibration or punning (rodding), a practice made necessary because *air-voids* will equally reduce the strength of concrete. Remember, as a guide, that every 1% of voids left in the concrete, whether water-voids or air-voids, results in a strength loss of about 5%. For example, 4% of voids would cause a strength loss of roughly 20%. Another drawback to a high water content is that it leads to increased *drying shrinkage*.

Water:cement ratio

It is usual to express the amount of water in any mix as a ratio of the weight of cement present. If this rule is applied to the case where a cement paste with the minimum water was used to fully hydrate the cement, the equation will be:

$$\text{Water:cement ratio} = \frac{\text{weight of water in mix}}{\text{weight of cement in mix}}$$

$$= \frac{0 \cdot 25 \text{ kgf}}{1 \text{ kgf}} = 0 \cdot 25$$

The range of water:cement ratios used in practice for ordinary dense concrete is between about 0·40 and 0·80.

Example 13 Find the total water required in a concrete mix specified to contain 100 kg of cement with a water:cement ratio of 0·50.

Re-arranging the water:cement ratio formula:

Weight of water in mix = water:cement ratio × weight of cement

$$= 0 \cdot 50 \times 100$$

$$= 50 \text{ kgf (or 50 l)}$$

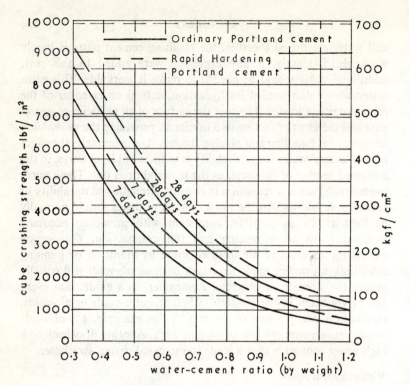

Fig. 22(a) Strength to water:cement ratio for concrete
(fully cured at 20°C)

An important relationship, called the *water:cement ratio 'law'*, is that the compressive strength of *fully compacted* concrete is governed by the water:cement ratio, irrespective of the mix proportions. This is illustrated by the graphs (Fig 22) for concrete made with ordinary and rapid-hardening Portland cements. These graphs only give *average* values obtained by testing a large number of specimens made with different cements, so results obtained using a particular cement may not be in exact agreement. A further provision is that the values only apply if the concrete is prevented from drying out (otherwise hydration could not continue) and where abnormal temperatures do not occur. The graphs relate to specimens cured in water at 20°C or 68°F. Of course, the strength of concrete increases with age. The graphs in Fig 22 apply to specimens tested 7 and 28 days after casting.

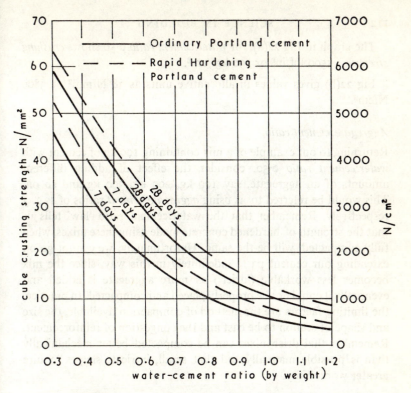

Fig. 22(b) Strength to water:cement ratio relationship for concrete
(fully cured at 20°C)

Type of cement

The rate of strength development of concrete varies with the type of cement used. Fig 35 shows typical strength curves for both ordinary and rapid-hardening Portland cement concretes, relative to a value of 100% at 28 days. You will see that the strength increases rapidly at first, but continues at a diminishing rate. In spite of the higher early strength of rapid-hardening Portland cement concrete, the ultimate strengths of the two types are about the same.

Example 14 Estimate from Fig 22 the 28 day cube strength of concrete of water:cement ratio 0·50, made with ordinary Portland cement, assuming full compaction and standard curing.

The graph in Fig 22(a), for *water:cement ratio* 0·50 shows *crushing strength* = 5500 lbf/in² or 390 kgf/cm².

Fig 22(b) gives values in alternative units as 38 N/mm² or 3800 N/cm².

Aggregate:cement ratio

Returning to our example of a mix containing 100 kg of cement with *water:cement ratio* 0·50, consider the effect of adding different amounts of an aggregate, say 100 kg, 200 kg, 300 kg and so on. This would be referred to as using *aggregate:cement* ratios of 1, 2, 3 respectively. Remember that the water:cement ratio 'law' tells us that the strength of hardened concrete made using these mixes, when fully compacted, will be the same. Unfortunately, we cannot go on extending our cement paste indefinitely in this way, since the mix becomes less workable ('drier') as more aggregate is added and eventually it could not be properly placed and compacted. In practice, the limiting factors are the method of compaction available, the size and shape of section to be cast and the congestion of reinforcement. Remember that drier mixes can be compacted better mechanically than is possible manually and that small, intricate shapes require greater workability.

Workability and strength

Considering again our concrete mix of 100 kg cement at a water: cement ratio of 0·50, suppose that we have decided that our aggregate: cement ratio is to be 4 and we use 400 kg of aggregate. This will fix the workability of the mix, but only relative to a particular aggregate, since the aggregate grading, particle shape and surface texture all influence the workability. The *finer* our aggregate, the *greater* its *specific surface* (the total surface area of all its particles per unit weight), and consequently the further must our cement paste be spread, with corresponding reduction in its lubricating efficiency, that is, *less workability*. This accounts for the use of a proportion of coarse aggregate in dense concrete. The larger particles give better workability for a given water:cement ratio and aggregate:cement ratio; *alternatively*, they allow higher strength (by reduced water:cement ratio) for a fixed aggregate:cement ratio and a given workability. This is clearly illustrated by the crushing strengths given in Fig 23 for

1:6 mixes of comparable workability using different maximum aggregate size. Notice that the mortar required a considerably greater water:cement ratio (it would vary for different sand gradings and workability) giving a much lower strength. Fig 24 shows values of

Mix by volume	Aggregate max. size		Water-cement ratio	Compressive strengths			
				7 days		28 days	
	in	mm		lbf/in²	N/mm²*	lbf/in²	N/mm²*
Concrete 1 : 6	1½	38	0·53	3400	23·5	5000	34·5
Concrete 1 : 6	¾	19	0·60	2700	18·5	4200	29
Concrete 1 : 6	⅜	9·5	0·70	2000	14	3200	22
Mortar 1 : 6	3/16	4·75	1·70	300	2	450	3

* For values in bar multiply by ten

Fig 23 Table of typical crushing strengths of 1 : 6 concrete and mortar mixes of medium workability with O.P. cement, irregular gravel and natural sand, showing the effect of particle size

Mix by volume	Aggregate max. size		Water-cement ratio	Drying shrinkage (per cent)		
				7 days	28 days	90 days
	in	mm				
Concrete 1 : 6	¾	19	0·60	0·005	0·015	0·03
Mortar 1 : 6	3/16	4·75	1·70	0·045	0·07	0·08
Mortar 1 : 3	3/16	4·75	0·90	0·055	0·09	0·11

Fig 24 Table of typical natural (in air) drying shrinkage of concrete and mortar mixes of medium workability

natural drying shrinkage for mixes of different maximum aggregate size, water:cement ratio, and cement content.

Aggregate for concrete

The maximum particle size of aggregates used in concrete is limited by the type of work concerned; for example, for reinforced concrete it is usually ¾ in (19 mm), due to the need for it to pass between the reinforcement.

Most structural concrete is made using *continuously graded* aggregates, that is, aggregates containing a suitable proportion of all particles sizes from the largest to the smallest. The particles of such aggregates will pack together, leaving a relatively small volume of voids to be filled by the cement paste. In practice, the main requirements can be met by using aggregates complying with BS grading limits for concrete aggregates (BS 882), and combining the materials in accordance with certain *nominal* or *standard* mix proportions, which will be given later in this chapter.

The British Standard makes provision for the following classes of concrete aggregate:

[a] Fine aggregate (e.g. natural and crushed stone sands)

[b] Graded coarse aggregate (continuous gradings), e.g. graded $1\frac{1}{2}$–$\frac{3}{16}$ in (38–4·76 mm) or graded $\frac{3}{4}$–$\frac{3}{16}$ in (19–4·76 mm)

[c] Single-size coarse aggregate (separate sizes of coarse aggregate) e.g. $\frac{3}{4}$–$\frac{3}{8}$ in (19–9·5 mm), or $\frac{3}{8}$–$\frac{3}{16}$ in (9·5–4·76 mm)

[d] All-in aggregate (coarse and fine aggregate intermixed—often as dug), e.g. $1\frac{1}{2}$ in (38 mm) down ballast

Note The metric equivalents given here are approximate.

The highest grades of concrete are made using two or more sizes of single-size coarse aggregate and a fine aggregate, each constituent aggregate being measured separately. In this way, close control can be maintained on the respective proportions of the different sizes of aggregate to achieve a good overall (combined) grading.

Good dense concrete may also be made by combining a fine aggregate with a graded coarse aggregate, since this allows a fair control of the overall grading, and especially the fines content, which is the most critical factor.

Finally, as an alternative to combining different classes of aggregate, an all-in aggregate can be used. However, this allows no direct control of the overall grading, which may prove unsatisfactory, particularly with regard to the ratio of fine to coarse material present. Attempts are sometimes made to overcome this objection by mechanically separating the fine and coarse constituents, then recombining them in suitable ratios, to give what is termed a *reconstituted* all-in aggregate. Nevertheless, all-in aggregate is not normally permitted to be used in high-class, or *structural* concrete, but is used in *foundation class* or *mass* (non-structural) concrete.

Aggregate grading

BS grading requirements for concrete aggregates are summarised in Tables 17, 18 and 19 (pages 312–13), which deal with fine aggregates, coarse aggregates (graded and single-sized) and all-in aggregates, respectively. Fine aggregates are classified into four *grading zones*, ranging from the coarsest concreting sands (Zone 1) to the finer varieties (Zone 4). Zone 4 sands are not normally permitted to be used in reinforced concrete unless it is shown by tests that the particular concrete is suitable for the work. This is merely a safeguard because of the extreme fineness of these sands, which can produce perfectly good concrete when the mix is proportioned accordingly (e.g. by mix design methods).

Proportioning concrete mixes

In many cases it may be impractical or uneconomical to resort to the use of mix design methods, in which case either *nominal* or *standard* mixes can be used.

Nominal mixes

These are designated in proportions *by volume*, based on dry material and using separate fine and coarse aggregates. In the case of foundation class concrete the equivalent 'all-in' mixes may be used. The basic nominal mixes are as follows (parts by volume).

[a] Reinforced concrete
  (i) 1:1:2 nominal (cement:fine:coarse)
  (ii) 1:1½:3 nominal (cement:fine:coarse)
  (iii) 1:2:4 nominal (cement:fine:coarse)
[b] Foundation class concrete
  (i) 1:2:4 nominal (cement:fine:coarse)
  or 1:5 (cement:all-in aggregate)
  (ii) 1:3:6 nominal (cement:fine:coarse)
  or 1:7 (cement:all-in aggregate)

Notice that the basic nominal mix proportions all have the ratio of fine to coarse aggregate 1:2 by volume. This is based on a voids ratio for the coarse aggregate of roughly 40%. The addition of sand will normally cause *particle interference* (that is, sand grains between coarse aggregate particles push them apart)—effectively increasing the voids ratio. Allowing 50% of sand normally ensures an adequate amount of mortar to fill the voids, which will prevent *honeycombing*.

The alternative mixes with *all-in* aggregate given above for foundation class concrete are in each case roughly equivalent to the nominal mix quoted. For example, with the 1:2:4 mix, 2 volumes of fine aggregate mixed with 4 volumes of coarse aggregate will combine to less than 6 volumes owing to the voids of the coarse aggregate. This is easily verified, and the equivalence of the 1:5 mix checked, by experiment.

In nominal mixes, the use of an invariable ratio of 1:2 for the fine to coarse aggregate will evidently not always result in the best overall grading being obtained. This fact is often recognised, and provided for, in specifications by permitting this ratio to be varied between the limits 1:1½ to 1:3, *provided* this results in a denser or more workable concrete, and that the sum of the volumes of fine and coarse aggregate for the particular nominal mix proportions is not altered. As an example of this, it should be permissible where a nominal 1:2:4 mix is specified to use say a 1:2½:3½ mix which will give a higher sand content, suitable with coarse sands, or a 1:1½:4½ mix which will give a lower sand content, suitable with fine sands. These are the extreme cases. Notice that the sum of the volumes of fine and coarse aggregate does not vary in either case, viz:

Proportions for nominal 1:2:4 mix	*Sum of aggregate volumes*
[a] 1:2:4	2 + 4 = 6
[b] 1:2½:3½	2½ + 3½ = 6
[c] 1:1½:4½	1½ + 4½ = 6

In addition to stating *mix proportions* of concrete, in specifications the required *minimum compressive strength* of test cubes prepared from the concrete as placed in the works is frequently given, and this effectively limits the maximum water:cement ratio for the mix, so that actually specifying a water:cement ratio, or water content, is unnecessary.

Fig 25 shows a typical form of steel mould used for casting test cubes. Since, for ordinary Portland cement concrete, the works cube tests are usually required to be made after curing for 7 or 28 days, the advantage of a preliminary or trial mix before the main site works are commenced is obvious. Typical strength requirements for dense concrete are given in Tables 20 and 21.

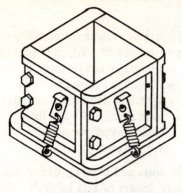

Fig 25 Steel mould used to cast concrete test cubes

Batching concrete mixes

The fact that a concrete mix may be designated or 'specified' by volume (e.g. nominal mixes) does not mean that the materials must necessarily be *batched*, that is, measured, by volume. If the weight per unit volume of the materials is known, it is quite a simple matter to convert volumes to their equivalent weights.

Example 15 A one bag *nominal mix* (i.e. using 1 bag of cement) of proportions 1:2:4 is to have a water:cement ratio of 0·50. Assuming *dry aggregates*, find the batch quantities and water content when batching [*a*] by volume [*b*] by weight. Take bulk densities as 100 lb/ft³ (1·60 kg/dm³) for sand and 90 lb/ft³ (1·43 kg/dm³) for coarse aggregate. Note that a 1 cwt bag of cement is taken as $1\frac{1}{4}$ ft³ equivalent bulk volume.

1 Using Imperial units and a 1 cwt bag of cement (taken as $1\frac{1}{4}$ ft³, based on a bulk density of 90 lb/ft³)

[*a*] *Volume batching*

   For proportions 1:2:4 by volume (cement:fine:coarse)

   Batch quantities = 1 bag:$2 \times 1\frac{1}{4}$ ft³:$4 \times 1\frac{1}{4}$ ft³

               = 1 bag:$2\frac{1}{2}$ ft³:5 ft³

   Water content   = 0·50 × 112 lb

               = 56 lb or 5·6 imperial gal

               Fine aggregate $2\frac{1}{2}$ ft³

               Coarse aggregate 5   ft³

               Water content $5\frac{1}{2}$ gal

[b] *Weight batching*

For proportions 1:2:4 by volume (cement:fine:coarse)

Batch quantities = 1 bag:2½ ft³:5 ft³, as in part (a)

= 1 bag:2½ × 100 lb:5 × 90 lb

= 1 bag:250 lb:450 lb

Water content = 5½ gal, as determined in part [a]

Fine aggregate   250 lb

Coarse aggregate 450 lb

Water content   5½ gal

2 Using the metric units and a 50 kg bag of cement (taken as 35 dm³, based on a bulk density of 1·43 kg/dm³)

[a] *Volume batching*

For proportions 1:2:4 by volume (cement:fine:coarse)

Batch quantities = 1 bag:2 × 35 dm³:4 × 35 dm³

= 1 bag:70 dm³:140 dm³

Water content = 0·50 × 50 kg

= 25 kg (25 litres)

Fine aggregate   70 dm³

Coarse aggregate 140 dm³

Water content   25 litres

[b] *Weight batching*

For proportions 1:2:4 by volume (cement:fine:coarse)

Batch quantities = 1 bag:70 dm³:140 dm³, as in [a]

= 1 bag:70 × 1·6 kg:140 × 1·43 kg

= 1 bag:112 kg: 200 kg

Water content = 25 l, as determined in [a]

Fine aggregate   112 kg

Coarse aggregate 200 kg

Water content   25 litres

In practice, aggregates used will often be damp, and this necessitates adjustment to the batch quantities:

(i) *For volume batching* Allowance must be made for *sand bulking*. Often, an average of 20% bulking is assumed, and this additional amount is added to the mix. Alternatively tests may be made to determine the actual percentage bulking of the sand, as described in Chapter 4. No bulking allowance is needed for coarse aggregate. For all-in aggregate an allowance of 5% to 10% may be made (based on 20% for sand alone, but taken pro-rata to the actual sand content) or alternatively a bulking test applied.

Adjustment of mix proportions is made as shown in the next example.

Example 16 Find the batch quantities for a one bag (50 kg) *nominal mix* of proportions 1:2:4 allowing for 20% bulking of fine aggregate.

For proportions 1:2:4 by volume (cement:fine:coarse)

Batch quantities, dry $= 1$ bag:2×35 dm³:4×35 dm³

$= 1$ bag:70 dm³:140 dm³

Batch quantities, damp $= 1$ bag:$70 \times \dfrac{120}{100}$ dm³:140 dm³

$= 1$ bag:84 dm³:140 dm³

Fine aggregate (bulked)   84 dm³

Coarse aggregate   140 dm³

(ii) *For weight batching* To allow for damp aggregates a greater weight of the aggregates must be measured and a corresponding reduction must be made in the water measured.

Example 17 The dry batch quantities for a concrete mix are specified as follows (taken from Example 15): sand 112 kg, gravel 200 kg, water 25 l. Calculate the batch quantities to be measured at the mixer for damp aggregates with moisture contents of: sand 5%, gravel 2%.

$$\text{Moisture in sand} = \frac{5}{100} \times 112 = 5 \cdot 6 \text{ kg}$$

$$\text{Moisture in gravel} = \frac{2}{100} \times 200 = 4 \cdot 0 \text{ kg}$$

Total moisture in aggregates   9·6 kg

Amount of sand to be measured $= 112 + 5 \cdot 6$

$= 117 \cdot 6$ kg

Amount of gravel to be measured $= 200 + 4 \cdot 0$

$= 204$ kg

Amount of water to be measured $= 25 - 9 \cdot 6$

$= 15 \cdot 4$ l

Batch quantities damp: sand 118 kg

gravel 204 kg

water 15·4 l

Standard mixes

These are mixes specified as dry *weights* of fine and coarse aggregate per unit weight of cement. Such mixes are tabulated in BSCP 114 as the weights of aggregate per unit weight of ordinary Portland cement. Data are given (see Table 21) for three standard mixes, with *minimum works cube strengths* at 28 days of 3000, 3750 and 4500 lbf/in² (or 21, 25·5 and 30 N/mm²). Different batch weights are given according to the degree of *workability* required (*low*, *medium* or *high*) and the maximum size of aggregate to be used ($1\frac{1}{2}$ in/38 mm, $\frac{3}{4}$ in/19 mm, $\frac{1}{2}$ in/13 mm or $\frac{3}{8}$ in/10 mm). The data relates to mixes using a Zone 2 sand to BS 882, but provision is made for the adjustment of material weights when using coarser (Zone 1), or finer (Zone 3) sands (Zone 4 sands are not applicable). A further adjustment is needed when using aggregates with specific gravities which differ significantly from 2·6. The adjustment is made on a pro-rata basis. The CP 114 data is based on *moderate* quality control in production for building works (defined statistically as a standard deviation not exceeding 1000 lbf/in² or 7 N/mm² at 28 days), and *weight batching* is specified. However, there is a proviso that mix proportions may be adjusted after a statistical analysis of the first 40 works cube test results, that is, where control is shown to be better than moderate by a lower standard deviation. For large works this offers a means of economising in the use of cement without resorting to a preliminary mix design and trial mixes. Standard mixes are also featured in BSCP 116 'The structural use of precast concrete' which gives data for use in works with a high degree of control (standard deviation not exceeding 500 lbf/in² or 3·5 N/mm²) as well as the same mix data as in CP 114. The *standard deviation* is a statistical measure used in mix production control of the *scatter* of a large number of test results about their *mean* (average) value. The better the control achieved, the less the scatter, and the lower the value for the standard deviation. An explanation of the calculation of standard deviations from test results is given in most books on applied mathematics and elementary statistics.

Designed mixes

Where the properties of the materials to be used are known, or are determined by tests, it is possible to design a concrete mix in order to achieve the required properties in the concrete, such as workability, strength and durability.

Designed mixes are provided for in BSCP 114 as an alternative to nominal or standard mixes, but stringent measures are specified for both the design and production stages where they are to be adopted.

Numerous methods of mix design are in use, but only a simple introduction to the subject, applicable to *dense concrete* of *medium strength*, is proposed here. For high strength, lightweight and 'dry lean' concretes different procedures are in use.

Most methods use the *crushing strength* of the hardened concrete as the main criterion, since it is a fairly good, and convenient, index of most of the important properties of dense concrete. Specifications for dense concrete with ordinary Portland cement usually give the minimum strengths required for works cubes tested at 7 and 28 days, and these are often used as the basis of a design. Remember the three important properties, *workability*, *strength* and *durability*, and the fact that workability is determined primarily by the *aggregate: cement ratio*; the strength and durability by the *water:cement* ratio. An outline procedure for mix design can now be given.

A concrete mix design procedure

1 Decide what *average strength* is necessary to ensure that no cube results fail to reach the minimum specified. In practice, 1 or 2% of failures are usually permitted on the grounds of statistical probability.

Average strength = specified minimum + margin of safety

The margin of safety or *control margin* needed depends on such factors as the degree of control at the works, method of batching (normally weight batching for designed mixes), and the efficiency of the mixing plant. Typical values with weight batching would be from ⅓ to ⅔ of the specified strength, or alternatively 2 to 2⅓ standard deviations, assuming the standard deviation is known or assessed. See Fig 26.

2 Get the *water:cement ratio* which corresponds to the average strength, from available data (Fig 22)

3 Check that this water:cement ratio is also low enough for the

Degree of control	Standard deviation		Control margin		Fraction of specified works cube strength
			Twice* standard deviation		
	lbf/in²	N/mm²†	lbf/in²	N/mm²†	
Moderate	1000	7·0	2000	14·0	$\frac{2}{3}$
Good	750	5·5	1500	11·0	$\frac{1}{2}$
Very good	500	3·5	1000	7·0	$\frac{1}{3}$

*this gives a statistical probability of not more than 2½% works cube test *failures* (2½ standard deviations would reduce this to 1% probable failures)
†multiply by ten for equivalent values in bar

Note Metric equivalents in this table have been rounded and are therefore only approximate.

Fig 26 Table of suggested control margins to be added to the minimum 28-day cube strength to obtain the design average strength for concrete to be batched by weight (two methods)

durability needed (see Fig 27). If not, use the lower water:cement ratio demanded for durability.

4 Select the *aggregate:cement ratio* from published data (e.g. Fig 28)[1] to give the leanest (least cement content) mix possible for the required workability, at the predetermined water:cement ratio. Fig 29 gives examples of typical uses for concrete of different workabilities.

5 Check the design by a *preliminary* (laboratory) mix, and adjust it if necessary. This normally entails checking the concrete for workability, adequate cohesion and surface finish in addition to testing cubes for compressive strength (e.g. 3 at 7 days and 3 at 28 days). *Note* A specification may require higher minimum strengths for these preliminary tests than for works cube tests.

6 Check the mix *full-scale*, using the actual plant for the works (the tests made are normally those listed in Section 5 above)

[1] Published with the permission of the Cement and Concrete Association (Ref. Table 6 of their publication Ebl, October 1958 edition: '*An introduction to cement and concrete—notes for students*')

Example 18 Prepare a mix design for a laboratory trial mix for dense concrete to the following specification, where site control is good, with weight batching.

Concrete workability Medium

Minimum works cube strength–3000 lbf/in² (211 kgf/cm², or 20·7 N/mm²) at 28 days with O.P. cement

Exposure—Normal outside, in temperate climate

Aggregates $\frac{3}{4}$–$\frac{3}{16}$ in (19–5 mm) graded crushed rock to BS 882 and natural sand, BS Zone 3

Design:

Average strength=3000+control margin
$$=3000+(\tfrac{1}{2}\times3000) \text{ for good control (see Fig 26)}$$
$$=4500 \text{ lbf/in}^2$$

Alternatively, using *metric* data

[*a*] With gravitational force units,
$$\text{Average strength}=211+(\tfrac{1}{2}\times211)=316 \text{ kgf/cm}^2$$

[*b*] With absolute force units,
$$\text{Average strength}=20\cdot7+(\tfrac{1}{2}\times20\cdot7)=31\cdot05 \text{ N/mm}^2$$

Water:cement ratio for average strength (from Fig 22)=0·57 (this applies for either the Imperial or metric data, and is within the limit of 0·60 for the degree of exposure—Fig 27)

Condition of exposure	Water:cement ratio*	
	Thin sections	Mass concrete
Moderate exposure	0·60	0·70
Severe exposure	0·55	0·65
Regular wetting and drying	0·50	0·60

* with loose production control reduce these values by 0·05

Fig 27 Table of maximum water:cement ratios for dense concrete under various conditions of exposure

Aggregate:cement ratio from Fig 28:

Water:cement ratio 0·50 relates to aggregate:cement ratio 4·0

Water:cement ratio 0·60 relates to aggregate:cement ratio 5·0

By interpolation:

water:cement ratio 0·57 relates to aggregate:cement ratio 4·7

| Workability | Water: cement ratio | Aggregate:cement ratio by weight | | | | | |
| | | Gravel aggregate max. aggregate size | | | Crushed rock max. aggregate size | | |
		1½ in 38 mm	¾ in 19 mm	⅜ in 9·5 mm	1½ in 38 mm	¾ in 19 mm	⅜ in 9·5 mm
Very low	0·4	5	4½	3½	4½	4	3
	0·5	7½	6½	5½	6½	5½	4½
	0·6	–	–	7½	–	7	6
	0·7	–	–	–	–	–	7
Low	0·4	4½	4	3	4	3½	–
	0·5	6½	5½	4½	5½	5	4
	0·6	7½	7	6	7	6	5
	0·7	–	8	7	8	7	6
Medium	0·4	4	3½	–	3½	3	–
	0·5	5½	5	4	5	4	3½
	0·6	7	6	5	6	5	4½
	0·7	8	7	6	7	6	5½
High	0·4	3½	3	–	3	3	–
	0·5	5	4	3½	4½	4	3
	0·6	6½	5	4½	5½	4½	4
	0·7	7½	6	5½	6½	5½	5

Fig 28 Table of aggregate:cement ratios for four degrees of concrete workability with different values of water:cement ratio. See also Fig 30 for sand contents

Sand content (see Fig 30)

For a Zone 3 sand and ¾ in (19 mm) maximum size coarse aggregate:

Sand content $= 30\%$

Parts of sand in 4·7 parts of total aggregate $= \dfrac{30}{100} \times 4\cdot7$

$= 1\cdot41$

Parts of crushed rock in 4·7 parts of total aggregate $= 4\cdot7 - 1\cdot41$

$= 3\cdot29$

Workability	Maximum aggregate size		Slump		Com-pacting factor	Suitable use of concrete
	in	mm	in	mm		
Very low	$\frac{3}{8}$	9·5	0	0	0·75	for simple reinforced sections with inten-sive vibration, or mass concrete with normal vibration
	$\frac{3}{4}$	19	$0-\frac{1}{2}$	0–15	0·78	
	$1\frac{1}{2}$	38	0–1	0–25	0·78	
Low	$\frac{3}{8}$	9·5	$0-\frac{1}{4}$	0–5	0·83	for simple reinforced sections with normal vibration, or mass concrete without vibration
	$\frac{3}{4}$	19	$\frac{1}{2}-1$	15–25	0·85	
	$1\frac{1}{2}$	38	1–2	25–50	0·85	
Medium	$\frac{3}{8}$	9·5	$\frac{1}{4}-1$	5–25	0·90	for hand placing and compaction without congested reinforce-ment, or in heavily reinforced sections with vibration
	$\frac{3}{4}$	19	1–2	25–50	0·92	
	$1\frac{1}{2}$	38	2–4	50–100	0·92	
High	$\frac{3}{8}$	9·5	1–4	25–100	0·95	for hand placing and compaction in congested reinforce-ment
	$\frac{3}{4}$	19	2–5	50–125	0·95	
	$1\frac{1}{2}$	38	4–7	100–175	0·95	

Fig 29 Workability of concrete related to conditions of placing and compaction

Coarse aggregate Max. size	Sand content*—percentage by weight of total aggregate for sand to BS 882 zone:			
	I	2	3	4
$\frac{3}{8}$ in (9·5 mm)	50	40	35	30
$\frac{3}{4}$ in (19 mm)	45	35	30	25
$1\frac{1}{2}$ in (38 mm)	35	30	25	22

*Data for this table relate to a 1:6 mix
Leaner mixes require a decreased coarse:fine ratio
Richer mixes may permit an increased coarse:fine ratio

Fig 30 Table of typical sand contents for concrete mixes

Mix proportions for trial mix (by weight)

1:4·7 (cement:aggregate), water:cement ratio 0·57
or 1:1·41:3·29 (cement:coarse:fine), W/C=0·57
usually written as 1:1·41:3·29/0·57

Adjustments to trial mixes

A mix designed by the simple method just used in Example 18 is more likely to require adjustment following the laboratory trial mix than a mix designed using more complex methods, but the end product should not be greatly different. If the mix has *insufficient cohesion* or is *harsh*, or *honeycombed*, the sand content should be increased, offsetting this by an equivalent reduction of coarse aggregate. If it is *oversanded* (showing the larger particles too widely dispersed) reduce the sand content. A mix which is too *workable* can be made leaner by increasing the aggregate:cement ratio; a mix which is too *stiff* should be made richer by decreasing the aggregate:cement ratio. The water:cement ratio should not be altered unless this is found necessary as a result of the strength tests on the preliminary test cubes. With standard curing (BS 1881 specifies curing under water at 20°C) this should only arise where the cement used is above or below the 'average' implied by the water:cement ratio/strength curves used (Fig 22). In these cases, it is important to allow for any known variability in the cement supply for the works.

EXPERIMENT 22 Compressive strength of different concrete mixes

Introduction Five mixes of different proportions by weight are to be prepared, using the least water to give a mix convenient for hand compaction. Fig 31 gives the amounts of materials for one cube (based on aggregates of s.g. 2·6). Good practice would require 3 cubes to be made and tested for each mix, to give an average result. If fewer cubes are made the results will be less reliable

Apparatus Compression testing machine—cube moulds (4 in/100 mm or 6 in/150 mm)—tamping bar (to BS 1881)—gauging trowels—finishing trowel—trays for mixing and waste—measuring cylinder (1000 and 500 ml)—physical balance (metric or imperial)—mould oil and brush—polythene film—waterproof crayon

Mix by weight		Amount for one cube*					
		6 in or 150 mm size			4 in or 100 mm size		
		Cement	Fine	Coarse	Cement	Fine	Coarse
1 : 1 : 2	kg	2·5	2·5	5·0	0·7	0·7	1·4
	lb	5·0	5·0	10·0	1·5	1·5	3·0
1 : 1½ : 3	kg	2·0	3·0	6·0	0·6	0·9	1·8
	lb	4·0	6·0	12·0	1·2	1·8	3·6
1 : 2 : 4	kg	1·5	3·0	6·0	0·4	0·8	1·6
	lb	3·0	6·0	12·0	0·9	1·8	3·6
1 : 2½ : 5	kg	1·2	3·0	6·0	0·4	1·0	2·0
	lb	2·4	6·0	12·0	0·8	2·0	4·0
1 : 3 : 6	kg	1·0	3·0	6·0	0·3	0·9	1·8
	lb	2·0	6·0	12·0	0·6	1·8	3·6

*amounts based on aggregate s.g. 2·6 will be more than required to fill the mould

Note Metric and Imperial quantities in this table are not exact equivalents as the figures have been rounded, but proportions are exact.

Fig 31 Quantities of cement and aggregates for concrete cubes

Materials O.P. cement: Fine and coarse aggregates (graded, maximum ¾ in/19 mm)—dry

Method (for one cube)
1 Weigh out the cement and aggregates for the first mix (Fig 31) and mix them thoroughly, dry
2 Mix a measured amount of water with the materials to achieve a uniform, workable concrete. Note the amount used
3 Oil the mould and baseplate, and check that all nuts are tight
4 Fill the concrete into the mould in 2 in (50 mm) layers, *compacting each thoroughly* with the tamping bar. Use *at least* 25 strokes per layer for a 4 in (100 mm) cube, or *at least* 35 strokes per layer for a 6 in (150 mm cube)
5 Level and smooth the top surface of the concrete flush with the edge of the mould
6 Repeat Sections 1 to 5 for the remaining mixes, marking the specimens to identify the mixes
7 Cover all moulds with polythene film and leave them for 24 hours

8 Mark each cube, demould it, then store it in water (at 20°C for standard tests) until tested in compression (wet) at 7 days or 28 days

Results Record these in a table, and also draw a graph to show the relationship between compressive strength and water:cement ratio (to be calculated) for the five mixes. Compare the graph with Fig 22 and comment on any difference.

YIELD OF MIXES

The volume yield of a concrete, mortar, or plaster mix can be calculated if the mix proportions and material specific gravities are known. It is taken to be the sum of their *absolute*, or *solid*, volumes, and assumes complete compaction without voids. In practice, compacted mixes normally contain 1 or 2% of air, but this is usually ignored in the calculation.

Example 19 Calculate the yield for the following concrete mixes if the specific gravities of the materials are: cement 3·15, sand 2·6, crushed stone 2·7 (water = 1·0) assuming full compaction.
[*a*] Mix 1. *Imperial units.* Batch quantities: cement 100 lb, sand 200 lb, stone 400 lb, water 60 lb (density of water 62·4 lb/ft³)

Material		Absolute volume (ft³)
Cement	$\dfrac{100 \text{ lb}}{3 \cdot 15 \times 62 \cdot 4} =$	0·51
Sand	$\dfrac{200 \text{ lb}}{2 \cdot 6 \times 62 \cdot 4} =$	1·23
Stone	$\dfrac{400 \text{ lb}}{2 \cdot 7 \times 62 \cdot 4} =$	2·37
Water	$\dfrac{60 \text{ lb}}{1 \times 62 \cdot 4} =$	0·96
	Total yield =	5·07 ft³

Note If the same mix had 6% of entrained air the yield would be increased to $5 \cdot 07 \times \dfrac{106}{100} = 5 \cdot 37 \text{ ft}^3$

[b] Mix 2. *Metric units*. Batch quantities: cement 50 kg, sand 100 kg, stone 200 kg, water 30 l (kg).

Material		Absolute volume (dm³, or l)
Cement	$\dfrac{50 \text{ kg}}{3\cdot15} =$	15·9
Sand	$\dfrac{100 \text{ kg}}{2\cdot6} =$	38·5
Stone	$\dfrac{200 \text{ kg}}{2\cdot7} =$	74·1
Water	$\dfrac{30 \text{ kg}}{1\cdot0} =$	30·0
	Total yield $=$	158·5 l
		(or 0·1585 m³)

Note If the same mix had 6% of entrained air, the yield would be

increased to $158\cdot5 \times \dfrac{106}{100}$

$$= 168 \text{ l } (0\cdot168 \text{ m}^3)$$

Example 20 A concrete mix has an aggregate:cement ratio of 6·0 (by weight), with a sand content of 30% and water:cement ratio 0·50. Calculate the density of the freshly mixed, fully compacted, concrete, if the specific gravities are: cement 3·15, sand 2·65 and gravel 2·6.

$$\text{Sand ratio} = \frac{30}{100} \times 6 = 1\cdot8 \text{ parts}$$

Coarse aggregate ratio $= 6 - 1\cdot8 = 4\cdot2$ parts
Mix proportions by weight $= 1:1\cdot8:4\cdot2/0\cdot50$
Total parts by weight $= 1 + 1\cdot8 + 4\cdot2 + 0\cdot5$
$\qquad\qquad\qquad\qquad = 7\cdot5$

Total parts by volume $= \dfrac{1}{3\cdot15} + \dfrac{1\cdot8}{2\cdot65} + \dfrac{4\cdot2}{2\cdot6} + \dfrac{0\cdot5}{1}$
(dividing by s.g.)
$\qquad\qquad\qquad\quad = 0\cdot32 + 0\cdot68 + 1\cdot62 + 0\cdot5$
$\qquad\qquad\qquad\quad = 3\cdot12$

Mean s.g. of mix $= \dfrac{7\cdot5}{3\cdot12} \quad = 2\cdot40$

E

$$\text{Density of mixed concrete} = 2\cdot 40 \text{ kg/l}$$
$$= \text{or } 2\cdot 40 \text{ tonne/m}^3$$

Note In imperial units, taking the density of water as 62·4 lb/ft^3:
density of mixed concrete $= 2\cdot 40 \times 62\cdot 4 = 150$ lb/ft^3

WORKABILITY TESTS

Workability refers to the ease with which concrete can be placed and compacted, and various tests are used to assess this.

Slump test

The most commonly used site method for assessing workability is the *slump* test, as described in BS 1881. The test entails filling an open-ended conical mould (Fig 32) with concrete and measuring the slump, or drop in level, of the concrete when the mould is lifted clear (Experiment 23 gives fuller details). Since the workability of a given concrete increases with the amount of water added, the slump test also provides a means of controlling the water content of successive batches of the same mix, provided there is no substantial change in the aggregate gradings. Remember that an increased water content causes a reduction in strength and durability, but that workability must be sufficient to allow full compaction.

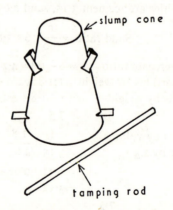

Fig 32 Slump test apparatus

Compacting factor test

Another test applied to workability uses the *compacting factor* apparatus (Fig 33), as specified in B.S. 1881. The extent to which concrete is compacted in falling through a fixed height is used as a measure of its workability (the method is given in Experiment 24). This test is particularly useful for mixes of low workability, at which the slump test becomes insensitive. Fig 29 gives examples of typical uses for concrete at different workabilities.

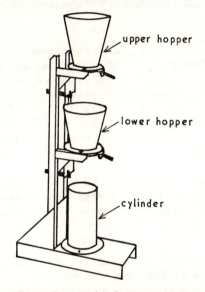

upper hopper

lower hopper

cylinder

Fig 33 Compacting factor apparatus

EXPERIMENT 23 Slump of concrete with varying water content
Note This experiment gives the standard method applicable to all dense concrete mixes. It is normally applied to the *freshly mixed* concrete, if in a laboratory 6 minutes after first adding water. In other cases the concrete should be remixed (no water or other materials being added) and the time after first adding mixing water reported with the result.

Apparatus Slump cone (Fig 32) and tamping rod (to BS 1881)—steel rule—straightedge—scoop—gauging trowels—steel float—mixing tray—measuring cylinder (1000 ml)—physical balance

Materials O.P. cement. Fine and coarse aggregates (graded, max. ¾ in/19 mm–dry)

Method The quantities of materials and water required for each slump test are given in Fig 34

1 Weigh out the required amounts of cement and aggregates for the first slump test and mix them thoroughly (dry)

2 Measure out the water required for the 0·50 water:cement ratio and mix it in to get a concrete of uniform consistency

Material	Amount for workability test* (slump or compacting factor) 1 : 2 : 4 mix by weight	
	kg	lb
Cement	2·5	5·0
Fine	5·0	10·0
Coarse	10·0	20·0
Water:	ml	ml
W/C=0·50	1250	1130
W/C=0·55	1370	1250
W/C=0·60	1500	1360
W/C=0·65	1620	1470
W/C=0·70	1750	1590

* material for one test based on aggregate s.g. 2·6

Note Metric and Imperial quantities in this table are not exact equivalents as the figures have all been rounded. The proportions are exact.

Fig 34 Materials for concrete workability tests (Experiments 23 and 24)

3 Check the slump cone for cleanliness and stand it on a solid, flat, impermeable and clean surface (the wider end as base)

4 Hold the cone firmly in position and fill it with concrete in four approximately equal layers, rodding each layer *exactly* 25 times. The final layer must slightly overfill the cone

5 Strike off surplus concrete from the cone and clear droppings from around its base

6 Lift the cone off steadily, vertically

7 Measure the slump (to the nearest $\frac{1}{4}$ in/5 mm) as the difference between the highest point on the slumped concrete and its original level in the cone. This can be done by inverting the empty cone alongside the slumped concrete, placing a straightedge across it, and measuring down from the straightedge with a rule or, in some cases, by direct reading from a scale provided

8 Repeat the procedure (Sections 1–7) for the mixes of different water content (Fig 34)

Note If the slump specimen collapses or shears off, a *true slump* is not measurable. If this occurs, repeat the test, and if a true slump is still not obtained measure the result in the normal way and record it as a *collapse* or *shear* slump

Results Record these in a table, showing the amount and type of slump (no slump is recorded as 'nil')

Supplementary work Prepare and crush cubes for the mixes of different water:cement ratio. (Results will comply with the water: cement ratio law only if all can be, and are, fully compacted— vibration might be required for the drier mixes).

EXPERIMENT 24 Compacting factor test on concrete

Note This is a standard test applicable to all dense concrete mixes. It is normally applied to the *freshly mixed* concrete, if in a laboratory 6 minutes after first adding water, in other cases the concrete should be remixed (no water or other materials being added) and the time after first adding mixing water reported with the result. The method only is given here, but the test could be applied to any mixes proportioned as in previous experiments.

Apparatus Standard compacting factor apparatus to BS 1881 (Fig 33)—steel rod, $\frac{5}{8}$in (16 mm) diameter, 24 in (610 mm) long with rounded end—two finishing trowels—scoop—scales (25 kg × 10 g or 56 lb × $\frac{1}{2}$oz)—concrete mixing equipment—a B.S. rod, or *vibrator* for the compacting factor cylinder (a vibrating hammer may be used) is also needed, unless the weight of fully compacted concrete to fill the cylinder is to be *calculated* (see Example 22)

Materials O.P. cement. Fine and coarse aggregate

Method

1 Prepare a concrete mix of known proportions (suitable amounts of materials are given in Fig 34)

2 Weigh the cylinder (empty), place it on the stand and cover it with the finishing trowels to keep droppings clear

3 Fill the top hopper with concrete, avoiding pre-compaction, then release the gate for the concrete to fall into the next hopper, assisted only if necessary by gentle rodding from the top

4 Remove the finishing trowels from the cylinder, then release the gate of the second hopper so that the concrete falls into it. Assist with the rod if necessary

5 With the finishing trowels remove any concrete above the top edge of the cylinder

6 Wipe clean the outside of the cylinder, then weigh it together with the contents. Deduct the weight of the cylinder to get the weight of the partially compacted concrete (w)

7 Empty the cylinder, then refill it in 2 in (50 mm) layers, thoroughly vibrating or rodding it so as to remove all air

8 Wipe clean the outside of the cylinder, then reweigh it and get the weight of the fully compacted concrete (W)

Result

$$\text{Compacting factor} = \frac{w}{W}$$

Example 21 Calculate the compacting factor of concrete from the following test results

Weight of partially compacted concrete to fill cylinder   10·80 kgf

Weight of fully compacted concrete to fill cylinder   13·50 kgf

$$\text{Compacting factor} = \frac{10·80}{13·50} = 0·80$$

Example 22 Calculate the weight of fully compacted concrete to fill a compacting factor cylinder of capacity 5·50 l when the mix proportions are 1:2:4 *by weight* (cement:sand:gravel), with water: cement ratio 0·50. The specific gravities are: cement 3·10, sand 2·70 and gravel 2·60.

Material	Parts by weight	s.g.	Parts by volume (absolute)
	W	G	$= \dfrac{W}{G}$
Cement	1	3·10	0·32
Sand	2	2·70	0·74
Gravel	4	2·60	1·54
Water	0·5	1·00	0·50
Total parts	7·5		3·10

Mean s.g. of mix $= \dfrac{7·5}{3·1} = 2·42$

Weight of fully compacted concrete to fill cylinder (kfg) = volume in litres × mean s.g.
= 5·50 × 2·42
= 13·3 kgf

Other workability tests

A number of other standardised tests are used to measure concrete workability.

The *Vebe consistometer* test is used mainly in laboratories, and is particularly useful for dealing with very stiff mixes. The time taken, in seconds, for a moulded cone of concrete (formed using a *slump test* cone) to reform to cylindrical shape in a vibrated container is measured. Other tests measure the *rate of penetration* of a metal ball or cone through a concrete sample, or the *flow characteristics* of a moulded sample, in each case when subjected to oscillation.

CURING

All concrete must be maintained in a moist condition during the early stages of hardening:
[a] to ensure that moisture is available to hydrate the cement
[b] to reduce the rate at which the initial drying shrinkage occurs, in order to minimise cracking
[c] because this reduces the permeability, and increases the durability, of the concrete.

Since the mixing water in concrete exceeds that needed for hydration, curing is most simply accomplished by preventing its evaporation from the concrete. The period over which curing is important is obviously related to the rate of strength gain obtained with efficient curing (see Fig 35). For ordinary Portland cement concrete the first 7 to 10 days are critical. For work externally in very damp weather, and internally in the humid conditions often found in buildings under construction, evaporation may be so slow that *natural curing* is adequate. In most cases this cannot be relied on, and positive measures should be taken, such as covering the concrete with *impermeable sheeting*, or spraying surfaces with a *membrane curing compound*. It is particularly important to give protection from direct sunlight and drying winds. Where special cements are used, the manufacturer's recommendations for curing should be followed. It is specially important that *high alumina cement concrete* should be kept thoroughly wet for the first 24 hours.

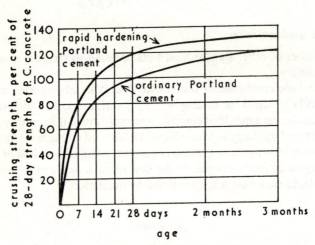

Fig 35 Rate of strength gain of concrete

Temperature effect

In common with most other chemical reactions, the rate of hydration of cement increases with temperature rise. The practical significance of this is that concrete hardens more slowly in *cold weather* (the data of Fig 35 apply for curing at 20°C).

Green (recently cast or unmatured) concrete must be protected from the disrupting effects which would accompany the freezing of its water content, and against collapse or damage by the premature removal of formwork or supports which might arise due to its slower rate of strength gain at low temperatures. Remember that, provided freezing does not occur within the concrete, the effect of low temperature on its strength is not permanent and the full strength will ultimately be obtained, but after a longer period.

During frosty weather but when *air temperatures* during the working day are *above freezing point*, the use of accelerators (e.g. calcium chloride added at 2 % of the weight of Portland cement), or of quick-setting or rapid-hardening (including extra rapid-hardening) cements, can assist, due to the increased rate at which the heat of hydration is given off and the more rapid strength gain.

It is essential that aggregates used should be free of frost, ice or snow, and stockpiles should therefore be covered.

When *air temperatures* fall *below freezing point* during the working day the additional measure of heating the materials before mixing is advisable. The water should be up to 60°C and the aggregates up to 25°C. In this case, mix the aggregates and water before adding the cement, to prevent a flash set.

None of the precautions taken can be fully effective unless they are supplemented by covering the work, and preferably insulating it to retain the heat of hydration of the cement.

PLACING

Concrete should be placed in the works as soon as possible after mixing, taking account of the setting characteristics of the type of cement used. Ordinary Portland cement concrete should normally be placed within about 30 minutes of mixing, although this period can be extended up to several hours without detriment to the material, provided it is kept *continuously agitated* until used and that loss of water by evaporation is prevented (this is because agitation suspends the setting action).

In gap-graded concrete the 'middle-sizes' of the aggregate are missing, with the result that the particle interference which occurs with continuous gradings is avoided. With the coarse particles touching, the mortar has merely to fill the voids between them. The amount of sand then needed is 5 to 10% less than with continuous grading. The exclusion of two, or three, of the middle sieve sizes of aggregate particle is required, for example by using any pair of the following combinations of coarse and fine aggregate:

Coarse aggregate	with	*fine aggregate*
$1\frac{1}{2}$–$\frac{3}{4}$ in (38–19 mm)		$\frac{3}{16}$ in –100 (4·75 mm–150 μm)
$1\frac{1}{2}$–$\frac{3}{8}$ in (38–9·5 mm)		No. 7 –100 (2·40 mm–150 μm)
$1\frac{1}{2}$–$\frac{3}{16}$ in (38–4·75 mm)		No. 14–100 (1·20 mm–150 μm)
$\frac{3}{4}$–$\frac{3}{8}$ in (19–9·5 mm)		No. 7 –100 (2·40 mm–150 μm)
$\frac{3}{4}$–$\frac{3}{16}$ in (19–4·75 mm)		No. 14–100 (1·20 mm–150 μm)
$\frac{3}{8}$–$\frac{3}{16}$ in (9·5–4·75 mm)		No. 14–100 (1·20 mm–150 μm)

DRY LEAN CONCRETE

The term *dry lean* is applied to concrete with an aggregate:cement ratio greater than 8, and in practice ratios of between 14 and 24 by weight are common. The minimum water content is used, to obtain a moist material suitable for rolling and used mainly as a road-base. A moisture content of roughly 5% (by weight of the dry materials) is typical. The sand content is usually 35 to 40%, and $1\frac{1}{2}$ in (38 mm) maximum aggregate is common.

LIGHTWEIGHT CONCRETE

The term lightweight was at one time reserved for concrete of density not exceeding 100 lb/ft³ (1·6 kg/dm³), but is now taken to include *no-fines* concrete, which has densities up to 120 lb/ft³ (1·92 kg/dm³) for aggregates of normal specific gravity.

Low density may be achieved in three ways:

[a] by omitting the fine aggregate (no-fines concrete)

[*b*] by the use of lightweight aggregates
[*c*] by aerating or foaming the concrete.

No-fines is made by mixing a coarse aggregate with cement and water. Aggregate may be normal or lightweight, and must be predominantly single-sized ($\frac{3}{4}$–$\frac{3}{8}$in/19–9·5 mm is usual). In effect, the particles of aggregate are held together at their points of contact by a neat cement paste. The water:cement ratio for this paste is critical, since if too stiff it will not coat the aggregate particles properly, and if too thin it will drain from the aggregate and accumulate at the bottom of the work. Highly absorbent aggregates are best hosed, then allowed to drain, before use. The mix, when using aggregates of normal specific gravity, is commonly 1:8 by volume, with a water:cement ratio of 0·40 or less. There must be no delay in placing the mixed concrete.

Apart from its thermal insulation properties when surfaces are sealed, no-fines concrete is not subject to capillarity, and in consequence can provide a weatherproof wall in one thickness (often 8 in/200 mm) if rendered externally. Cube strengths of 1000 lbf/in² (70 kgf/mm², or 6·9 N/mm²) and above at 28 days with ordinary Portland cement are common.

Lightweight aggregate concrete Typical examples are concrete made with clinker aggregate or foamed blast-furnace slag. A large part of the production of lightweight concrete is in the form of precast concrete blocks or slabs, with mix proportions by volume from about 1:6 to 1:10 (cement:aggregate). These are used mainly for internal partitions, and are classed either as loadbearing or non-loadbearing, according to strength and thickness.

Cellular (aerated or foamed) concrete No coarse aggregate is used in these 'concretes', and for the lowest density products the fine aggregate also is omitted, in which case lime may be added. The principle is for large numbers of minute air or gas bubbles to be dispersed throughout the concrete, giving reduced density and strength. By controlling the density, a concrete which has the best possible thermal insulating properties for a given strength may be obtained. To reduce the relatively high drying shrinkage, and to obtain increased strength, these products, particularly those containing lime, are frequently autoclaved (i.e. subjected to pressurised steam curing).

The simplest method of foaming concrete is to add to it a small quantity of a proprietary liquid foaming agent at the mixer. Hand mixing is unsuitable. However, to achieve the lowest densities the foam must be pre-formed in a special mixing chamber, then added as a separate ingredient. Alternatively a gas-generating additive, such as aluminium powder (0·2% by weight of the cement), can be added to the mix to cause the wet mix to swell to a cellular structure.

AIR-ENTRAINED AND PLASTICISED CONCRETE

These forms of concrete are made by introducing an additive, either at the mixer (usually a liquid) or incorporated in the cement. *Air-entrained* concrete is different from aerated or foamed concrete by the fact that coarse aggregate is not normally omitted. Also it has only a small content of entrained air, about 3 to 6%, and is therefore not a lightweight concrete. Unlike normal air or water-voids, entrained air is in the form of minute isolated bubbles, which do not lead to reduced durability; in fact the concrete is renowned for its frost-resistance. Although the entrained air itself causes a loss of strength, this is in most cases offset by the use of a lower water:cement ratio, made possible by the improved workability with entrained air.

Plasticised concrete has better workability than normal dense concrete, due to either reduced surface tension of the mixing water by an additive, or to the effects of adding a finely divided powder (e.g. lime or silica). Some plasticisers also entrain a minute amount of air.

Measuring air content

The air content of plasticised concrete may be determined by the *gravimetric method* outlined for mortars in Experiment 21 and Example 12, provided the mix proportions are known. The wet density of the mix is conveniently measured using the cylinder of the compacting factor apparatus.

Another common method, suited to field and laboratory tests, uses a pressure type air-meter of the type shown in Fig 36. A volume of normally compacted concrete is compressed inside an airtight container using a simple hand pump. This causes a reduction in volume of the air content in accordance with Boyle's law, and commercial air-meters are calibrated to allow a direct reading of the percentage air content when a specified pressure is applied. The method is not generally suitable where highly porous aggregates are used but is otherwise capable of giving reliable results. The method of use is given in the next experiment.

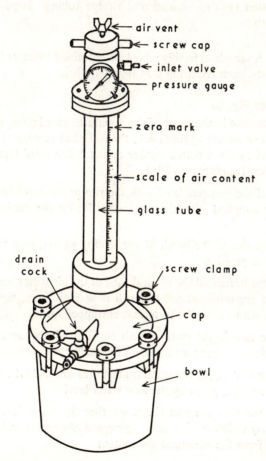

Fig 36 Air-meter used to find the air content of concrete (Experiment 25)

EXPERIMENT 25 Air content of concrete by air-meter

Note The following procedure is applicable to a typical commercial apparatus designed to take a sample of 0·25 ft³ (7·1 l) and correctly calibrated to give a direct reading of air content for an applied pressure of 15 lbf/in² (1·055 kgf/cm² or 10·34 N/cm²).

Apparatus Air-meter, with pump—sample container—scoop—steel rod ⅝ in (16 mm) diameter, 24 in (600 mm) long with rounded end—rubber mallet (250 g)—funnel and rubber tubing—large beaker—drying cloth

Specimen A sample of freshly mixed air-entrained concrete (sufficient for two tests, each requiring 0·25 ft³/7·1 l)

Method (see Fig 36)

1 Place concrete in the bowl of the apparatus in 3 layers, each time tamping the concrete 25 times with the rod, then tapping the outside of the bowl 15 times with a rubber mallet.* The third layer should leave the bowl slightly overfilled

2 Strike off the concrete level with the rim of the bowl and lay the metal disc supplied with the apparatus flat on the surface of the concrete

3 Clean the rim of the bowl, fit the sealing gasket, then clamp the cap firmly in position

4 Add water to half fill the vertical tube of the air-meter (vent *open*), incline the apparatus at 30° and roll it several times, tapping the cap lightly with the mallet to release entrapped air

5 With the tube in the vertical position fill it with water and drain water off down to the zero mark on the scale

6 Close the vent, apply the standard test pressure and take a reading of the air content, given by the new water level

7 Release the pressure and check whether the water level returns to zero. Any difference is due to *aggregate absorption* and must be *subtracted* from the measured air content

8 Clean out the apparatus

9 Make a further test (Sections 1–8) on a fresh sample.
* As an alternative to Section 1 the concrete may be compacted by vibration

Result
The total apparent air content given by a reading obtained for Section 6 must be corrected for the aggregate water absorption obtained in Section 7. This still does not take account of contraction of absorbed air in the aggregate, often between 0·2 and 1 % for normal density aggregates, and if required the appropriate *aggregate correction factor* may be obtained and *deducted* from the result. This factor is easily obtained by a separate test using the aggregates only, in the same quantities as in the concrete sample.

Obtain the average of the two corrected test results.

Note Concrete without entrained air can normally only be compacted to about 98 %, so from a measured air content 2 % is usually deducted to obtain the amount entrained. Alternatively, a test on an unplasticised mix would give the amount to be deducted.

GRANOLITHIC CONCRETE

This is dense concrete specially made to resist wear and abrasion, especially for floor surfaces. Rich mixes of low water content are required (e.g. 1:1:2 by weight with W/C=0·40 or less). The coarse aggregate must be specially hard and strong with good particle shape (not flaky or elongated) and free of excessive silt or very fine particles. Suitable types include basalt, gabbro, hornfels, some limestones, porphyry and quartzite (see also BS 1201, and Table 10, page 307). Fine aggregate may be natural sand, crushed gravel, or crushed stone of one of the types suitable as coarse aggregates. The methods of laying and finishing granolithic concrete follow those for thick floor screeds (see Chapter 6).

CONCRETE ROOFING TILES

These are obtainable in the majority of shapes and sizes applicable to single lap (interlocking) and double lap clay tiles (see Chapter 9) as well as in numerous special designs of interlocking tile. Those which have a flat upper surface are also commonly referred to as concrete *slates*.

The raw materials are sand, Portland cement and pigments, or coloured cement. Manufacture is now automated, with a production cycle of only 24 hours from mixing to stacking. After thoroughly mixing the materials with water the resulting mortar is extruded by the tile machine in a continuous ribbon onto a line of metal pallets shaped to form the underside of the tile. The upper surface is shaped by machine. The tiles are mechanically trimmed to length, nail holes are punched, and the tile is then usually surfaced with burnt sand (sometimes mixed with pigments and cement). The palleted tiles are

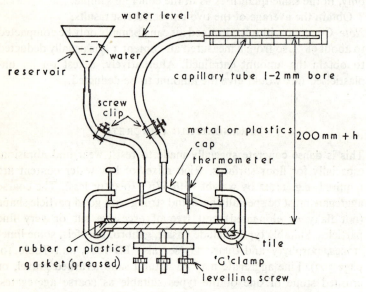

h = height to which water will stand in the
     capillary tube due to capillarity alone

Note on cap   An alternative to the use of the
   clamps and gasket is to attach the cap to
   the tile using Faraday wax, sealing the edges
   of the cap and covering the exposed tile
   surface at the top and edges

Fig 37 Permeability test on a concrete tile

stored in a warm, humidified chamber until the following day, when they are ready for stacking.

BS 473 'Specification for concrete roofing tiles and fittings' states typical dimensions for both double lap (Group A) and single lap (Group B) products and gives the permissible variation on size. The requirements for camber (where present) and thickness are given, also those for nail holes and nibs. There is a *transverse strength* test, similar to that for clay plain tiles (Fig 43) and a *permeability* test, for which a typical arrangement is shown in Fig 37. For the permeability test, tiles are first air-dried at room temperature. After sealing them into a metal or plastics cap, either by making a waxed joint, or by clamping against a rubber or plastics gasket, they are subjected to a head of water of 200 mm for 24 hours. The rate of flow of water through the tile, for the same head, is then measured. The BS limits for these tests are given in Table 22, page 314.

CONCRETE PIPES

Concrete pipes are made in various lengths and diameters for surface water and for soil (sewage); see BS 556 'Specification for concrete cylindrical pipes and fittings including manholes, inspection chambers and street gullies'. Pipes made using sulphate resisting or high alumina cement in addition to pipes made with ordinary Portland are available (both are resistant to sulphates, and high alumina cement resists acids). Some pipes are centrifugally *spun* in manufacture to remove free water and densify the concrete and others are reinforced to give extra strength.

A grade of *porous concrete pipe* is very popular for use in land drainage (BS 1194 'Concrete porous pipes for under-drainage').

CONCRETE BRICKS AND BLOCKS

These are dealt with in the next chapter.

Summary

Concrete is an artificial rock made using cement as binder together with various aggregates such as sand, gravel, crushed rock and various lightweight and special granular materials.

Dense structural concrete is usually made using Portland or other cement, together with separate fine and coarse aggregates combined in proportions to give a suitable overall grading for maximum density and strength of the finished concrete. The alternatives are continuous or gap grading.

All-in aggregate is used mainly for non-structural (foundation class) concrete.

The strength of concrete is governed primarily by the water: cement ratio and degree of compaction of the mix.

$$\text{Water:cement ratio} = \frac{\text{weight of water in mix}}{\text{weight of cement in mix}}$$

Compression tests made on works test cubes are normally at 7 and 28 days for ordinary Portland cement concrete.

The workability of concrete is governed primarily by the aggregate: cement ratio. Tests include the slump and compacting factor methods.

$$\text{Compacting factor} = \frac{\text{weight of partially compacted concrete to fill cylinder}}{\text{weight of fully compacted concrete to fill cylinder}}$$

For a given water:cement ratio and workability, leaner mixes (greater aggregate:cement ratios) apply for aggregates of low specific surface.

For the highest strengths a very low water:cement ratio is necessary, which, to allow full compaction, is only possible with rich mixes (low aggregate:cement ratios).

Low water:cement ratio and full compaction are consistent with good durability of concrete.

Air-entrainment results in improved frost resistance.

In practice, strength must be considered in relation to weight, thermal insulation and cost.

Structural concrete mixes may be specified as nominal mixes (by volume), standard mixes (by weight) or design mixes (usually by weight). Batching of standard and designed mixes is normally required to be by weight. Batching of nominal mixes may be by volume, but is preferable by weight.

When volume batching allow for bulking of aggregates.

When weight batching adjust the batch quantities, including water measured, to allow for any moisture in aggregates.

A typical sequence for dense concrete mix design procedure is:

1 Add to the specified minimum strength the appropriate control margin, to obtain the required average strength
2 Determine the water:cement ratio which corresponds to the required average strength
3 Check the water:cement ratio needed for durability. Select the lower of the two values considered (from Sections 2 and 3)
4 Choose the aggregate:cement ratio to give adequate workability
5 Establish the percentage weight of fine to total aggregate for the aggregates available
6 Make a trial mix, carry out tests and adjust mix proportions if necessary
7 After final adjustments, re-check the mix, then try a full-scale mix using the works plant

Curing the concrete ensures proper hydration (and strength gain) of cement, suppresses drying shrinkage and reduces permeability

At low temperatures hydration (and strength gain) is slowed down, but the ultimate strength should be the same provided that actual freezing of concrete and premature loading are avoided.

Cold weather precautions include the use of rapid-hardening or extra rapid-hardening cements and the covering of aggregates and the finished work. In extreme conditions materials for mixing may be pre-heated.

Dry lean concretes of mix proportions about 1:20 are used mainly for rolling to form road bases.

Lightweight concretes are made either by using lightweight aggregates, by omitting fine aggregate (no-fines) or aerating or foaming mortars. They may be used in-situ or precast.

Precast products may be steam-cured to improve strength and reduce drying shrinkage.

Granolithic concrete is a rich, dry mix containing selected hard aggregates, used as an abrasion-resistant finish to floors.

Concrete roofing tiles are available in many sizes and shapes, including single-lap and double-lap varieties. They may be pigmented and sand-faced. BS tests include transverse strength and permeability.

Concrete pipes are made in many lengths and diameters and may be spun or reinforced for extra strength. Sulphate resisting and high alumina cements may be used.

9

BRICKS, BLOCKS AND CLAY PRODUCTS

CLAY AS A RAW MATERIAL

Clay is commercially important as a raw material because it occurs naturally, can be brought to a plastic condition for moulding, and can then be dried and burnt, or *fired*, to convert it to a hard, durable material of fixed shape. The products are known as clay *ware*, clay *products* or *ceramics*.

Clays

All clays were formed by the weathering and disintegration of primary (igneous) rocks, and their composition varies. Clays are composed mainly of silica and alumina (aluminium oxide), together with other compounds, such as iron oxides, lime, magnesia and water. In raw clay, the alumina is usually combined chemically with the silica as hydrated aluminium silicates, but silica may also be present in clay as intermixed sand. Clays containing large amounts of such sand are termed *loams*. Some other clays, containing chalk in quantity, are referred to as *marls*. The mineral *kaolin* (china clay) is a clay consisting almost wholly of pure hydrated aluminium silicate, and is useful for its special properties, for example, as a fire-resisting cement. Because of its white colour, it is also used in the manufacture of white Portland cement. *Bauxite* is the name given to clays rich in hydrated aluminium oxide, well known as an ore of the metal aluminium and as a raw material of aluminous cements.

A *plastic* or *pure* clay is one containing a high proportion of alumina, and these include the more workable raw clays. The texture of clays ranges from the more workable or plastic types to the very hard, laminated clays termed *shales*, which approach

149

the character of slates. Slates themselves are a highly densified form of clay.

The wide variation in the composition and texture of raw clays leads to corresponding variations in the physical properties of fired clay, and the nature of the clay itself governs the types of product for which it is suitable.

Firing

Silica alone will not melt except at exceedingly high temperatures, but together with alumina and in the presence of a flux, such as lime or iron oxide, it will fuse at much lower temperatures. The firing of the raw clay entails heating it to at least the point of incipient fusion, which varies with the clay's composition but is well over 1000°C.

Plastic clays are not generally as suitable for firing at such high temperatures as clays of high silica content, because the high shrinkage of the alumina gives them a tendency to warp and crack. As a result, plastic clays in general give a light, porous but strong product, quite suitable for most classes of bricks, blocks and roofing tiles.

The higher temperatures at which clays of *high silica content* can be burnt will produce the semi-vitrification and high density usually associated with *engineering bricks*, *unglazed clay pipes*, *quarry tiles*, *terra-cotta* and *faience*. It is also found in some roofing tiles. However, the best clays in this category are those which are oldest in the earth's history, since they are less liable to warp and crack during firing. Such clays are found near the coal measures of Great Britain.

Clays which can be satisfactorily fired at very high temperatures are, of course, used in the making of *firebricks*, which are available to resist temperatures ranging from about 1200 to 1800°C. Products which can withstand these temperatures are called *refractories*, and suitable clays are known as *fireclays*. Their silica content ranges from about 70% for average refractory bricks to about 95% for bricks of the highest refactoriness, known as *silica bricks*.

Products which are underfired, in terms of temperature or time, are liable not to be durable, especially if used externally and in exposed situations.

Iron oxides, although their total extent in clays is only a few per cent, have an important influence on the colour of fired clay apart from acting as a flux. This may vary from pink, through red to blue, in the order of increasing maximum firing temperature, but depending also on the extent to which oxidation can occur under the particular kiln conditions. A yellow colour is usually associated with *magnesia*.

The influence of *limestone*, or *chalk*, as a flux has already been mentioned. However, this should not be present as lumps, for their conversion to quicklime during firing would constitute unsoundness in the finished product and might lead to lime-blowing at surfaces.

Expansion of clay products

After leaving the kiln, clay products take up moisture from the atmosphere and undergo slight expansion in the process. This action may continue, at a decreasing rate, over a number of years before the equilibrium condition is reached, during which the total expansion may be up to o·1 % or more. Since the greater part of this long-term expansion normally occurs during the first ten days after leaving the kiln, it is advisable not to lay bricks or other clay products until after this period. This expansion, which is regarded as irreversible, is distinct from the very slight moisture movement which can occur at any time.

CLAY BRICKS

Brick classification

Bricks may be classified in relation to their use, namely, as engineering bricks, facing bricks, common bricks and specials.

Engineering bricks are used mainly for their structural value and impermeability, and the essential properties are a high compressive strength, together with a very low water absorption. Associated characteristics are a high density and some degree of vitrification. Well-known examples are *Staffordshire blue*, *Accrington red* and *Southwater*, named after their place of origin.

Facing bricks are bricks of good appearance or architectural value. They include *sand-faced* and *rustic* varieties, and some of the better *stock* bricks.

Common bricks are those which cannot be classified as engineering or facing bricks. They include bricks for internal, or external, use where neither high strength nor low water-absorption is essential, and where appearance is relatively unimportant.

Special bricks are those made to serve special purposes; for example, firebricks, soft 'rubbers', glazed bricks and paving bricks. The term is also applied to bricks made to non-standard size or shape, and to bricks made from different materials (e.g. sandlime, concrete and glass bricks; also clinker fixing bricks).

Manufacture of clay bricks

Many natural clays are suitable for brickmaking without the addition of other materials, but some require blending to achieve a suitable composition.

Clay bricks are moulded into shape while they are in the raw, plastic condition. They are then fired. During firing, certain physical and chemical changes occur which cause the brick to solidify to its final shape. Substantial shrinkage of the brick occurs during the burning, and this must be allowed for when moulding in order to achieve the desired finished size. Finished dimensions must conform to certain standard sizes and tolerances so that the bricks can be laid to the correct levels and with regular joints.

The sequence of operations used in preparing and processing the raw clay varies according to its composition and hardness. Any of the following stages may be included but not necessarily in the order given.

1 Clay digging (usually by mechanical methods)
2 Weathering (to break down the clay)
3 Washing and screening
4 Grinding (for hard clays and shales)
5 Blending (addition of chalk, lime, sand, breeze, etc.)
6 Tempering (addition of water and kneading to the required consistency)
7 Moulding (shaping) or extruding. This stage may include the sanding of one or more brick surfaces to produce a sand-faced brick
8 Drying
9 Burning

The *weathering* stage is often omitted entirely, as it is nowadays only used for some of the best quality facing bricks. In the early brick-yards, weathering throughout the winter invariably followed digging in the autumn. In some cases, blending preceded the weathering.

Moulding is done either by hand or by machine. Machine methods are generally more economical, except where special shapes are required, but hand-made bricks are sometimes preferred architec-turally. Developments in machine methods have brought in many attractive *rustic* (patterned surface) treatments.

Machine-made bricks may either be of the wirecut or the pressed variety. *Wirecut* bricks are made from fairly soft, plastic clays by moulding a block of sufficient size to allow it to be cut by wires into three or more brick units ready for drying. An alternative to mould-ing is to extrude a continuous length of clay, brick or block section, and cut it to size by wires. Wirecut bricks are usually without *frogs* which are indentations in the bed face of a brick, and may then usually be identified by the wire-marks left on them. Some wirecut bricks are re-pressed during the moulding stage, or after an interval to allow partial drying, to form a shallow frog. This extra compaction hardens the brick and can improve its durability. *Pressed* bricks are made from stiff clays and shales, moulded under high pressure, often in two or more stages; the first stage produces a flush-faced brick unit of rectangular section, the other stages form one or two frogs which result in additional compaction.

A drying stage precedes the burning stage in cases where the raw brick contains considerable water initially, and aims to reduce the tendency for distortion and cracking in the firing stage. The drying is essentially evaporation, either in a normal atmosphere or with controlled temperature and humidity.

Burning is done either in clamps or kilns. For *clamp burning*, fuel usually in the form of coke breeze, is either added to the clay before moulding, or may be laid between the layers of bricks forming the clamp. The burning of the clamp is started at special fire holes and continues for several weeks or more. The bricks are variable in quality due to different degrees of burning throughout the clamp, and some are underburnt or overburnt. For this and other reasons, clamp burning has been largely superseded by kiln burning, the exceptions being a few smaller works.

For *kiln burning* the bricks are stacked and burnt in either specially built chambers, or tunnels, with provision for controlled burning at regulated temperatures. The fuels which are used may be pulverised coal, oil or gas, fed into the kiln as required through special inlets.

CLAY BLOCKS

Blocks constitute larger units for walling than bricks, and are also used in floor and roof construction as filling, usually in conjunction with in-situ reinforced concrete. Their large size allows speedier erection of walls or partitions than is normally possible with standard bricks. In addition to this, less mortar is used. They are dimensioned to allow bonding with standard bricks, and have a very low moisture movement in comparison to normally cured concrete products. Clay blocks are normally made with *perforations* (small holes) or *cavities* (large holes), and are known as *hollow clay blocks*. This avoids undue distortion in drying and firing, which would occur by non-uniform shrinkage of a large solid mass of clay. It also reduces weight, thereby saving raw material which in turn makes handling and laying easier and improves thermal insulation.

The raw materials and manufacture of clay blocks are basically the same as for clay bricks, although hollow blocks are invariably extruded and cut to length by a wire. In the extrusion, one or more surfaces may be grooved to provide a key for plaster. Extrusion necessitates a well-mixed, plastic clay, free of lumps and preferably vacuumised to reduce air-voids. Firing may take place in either compartment or tunnel kilns.

BS REQUIREMENTS FOR CLAY BRICKS AND BLOCKS

General requirements, methods of sampling and tests for clay bricks and blocks (solid, perforated, hollow or cavity) are given in BS 3921 'Specification for bricks and blocks of fired brick-earth, clay or shale'. This defines *common, facing* and *engineering* types of brick and block for walling in general terms, and likewise defines three qualities of brick—*special quality* (durable under exposed conditions), *ordinary quality* (durable externally if sufficiently protected by the use of sound constructional methods) and *internal quality* (for internal use only). These categories are given clearer definition by certain BS test limits applied to them. For example, the *dimensions* and *minimum strength* of all units are the subject of specific requirements, and all blocks must comply in respect of tests for *squareness, bowing* and *twisting*. In addition, for engineering bricks there are maximum limits for *water absorption*, and for all facing and common bricks there is a maximum limit for *liability to efflorescence*. A limit for *soluble salts* content is given for facing and common bricks and blocks of special quality only. A summary of the BS tests follows, and particulars of BS limits applicable are given in Tables 23 to 27, pages 315–17.

1 Format

Bricks and blocks are designated by the nominal dimensions; their actual dimensions (also specified) are less, to allow for the thickness of a mortar joint (taken as $\frac{3}{8}$ in/10 mm).

2 Dimensional tolerances

In the case of bricks the total length, width and height of 24 units together is determined (see Table 23, page 315), whereas blocks are individually measured. Blocks are additionally tested for squareness, bowing and twisting.

3 Compressive strength

Units are tested wet in a compression testing machine. *Bricks* are tested between 3 mm plywood sheets (Fig 38). In the case of bricks with frogs, where these are to be laid upwards in the work, they are filled before testing with a cement mortar, which must harden to between 4000 and 6000 lbf/in^2 (281–422 kgf/cm^2) or 28–42 N/mm^2 before testing.

Blocks are faced with a layer of cement mortar covering the test bed faces, except blocks (or bricks) which are to be laid in the work using a divided joint. These are similarly tested on two parallel mortar strips $\frac{3}{8}$ in (9·5 mm) thick. Blocks for walls and partitions are tested with the load applied perpendicular to the normal bed faces. The area for calculating the crushing strength is the overall area of one bed face (the smaller of the two is taken). Blocks for structural floors and roofs are tested with the load applied perpendicular to the ends in which the cavities appear. The area for calculating the crushing strength is then the net area obtained by subtracting the

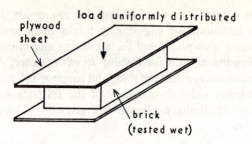

Fig 38 Crushing test on a brick

area of the cavities from the overall area of the end showing the cavities.

BS requirements for strength for engineering bricks are given in Table 24, page 315, and BS requirements for strength for bricks and blocks generally is given in Table 25, page 316.

4 Water absorption

Standard tests are made to determine the water absorption of oven-dried specimens after 5 hours boiling (or vacuumisation). There is also a 24 hour cold immersion test for works control purposes only.

BS limits are specified for engineering bricks (Table 24, page 315).

5 Soluble salts content

A chemical analysis is made of powder taken from 10 bricks or blocks, to establish the total amount of water-soluble salts and their composition (see Table 26, page 316).

6 Efflorescence test

The specimen is put in a close-fitting plastics bag, with a surface to be exposed in the work uppermost and uncovered. The specimen is saturated with distilled water and left in a warm, ventilated room to dry out. A further application of distilled water is then made. The brick is examined for efflorescence after it dries out, and the result is classified as *nil*, *slight*, *moderate*, *heavy* or *serious*, according to the extent of area affected, and the severity (see Table 27, page 317).

EXPERIMENT 26 Dimensions of clay building bricks

Apparatus Steel measuring tape

Specimens 24 bricks (a representative sample from the batch)

Method
1 Remove any small blisters or other small projections, and loose particles from the brick surfaces
2 Lay the bricks on a horizontal surface, butted together lengthwise (see Fig 39a):
either [a] in one continuous line of 24 bricks
   or [b] in two rows, each of 12 bricks
   or [c] in three rows, each of 8 bricks
         according to the length of tape available
3 Measure the overall length of the 24 bricks (adding the separate results together for procedure under Section 2[b] or 2[c])
4 Rearrange the bricks, butting them side by side, as in Fig 39b, and measure the overall width
5 Rearrange the bricks, butting the bed faces together, as in Fig 39c, and measure the overall depth

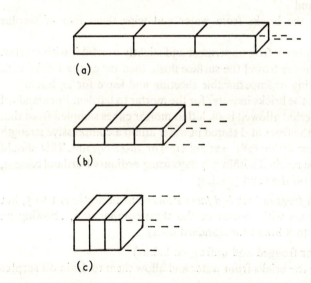

(a)

(b)

(c)

Fig 39 Checking dimensions of clay building bricks (Experiments 26)

Result
Check these (for Sections 3, 4 and 5) against the BS limits.

EXPERIMENT 27 Compressive strength of clay building bricks

Apparatus Compression testing machine (250 tonf /250 tonnef, or 2500kN)—steel rule—cloth. *For frogged bricks only*, materials for mortar and mixing equipment may be needed—see *Preparation* [*b*]

Specimens Clay building bricks

Materials Sheets of 3 mm plywood, rectangular 10×5 in (250× 130 mm)

Preparation
[*a*] *Bricks without frogs* Immerse in water (24 hours) before testing
[*b*] *Bricks with frogs* If frogs are to be laid upwards in the construction fill them with mortar before testing. If frogs are to be laid downwards no filling is required. Procedure for filling: *Bricks with frog in one bed face only*
1 Soak the bricks in water (24 hours for standard tests)
2 Prepare a stiff cement mortar, using 1 part Portland cement to 1½ parts sand
3 Remove the bricks from water and wipe them free of surplus moisture
4 Lay each brick frog uppermost and slightly overfill it with mortar. After 2–3 hours trowel the surface flush, then cover the bricks with damp sacking or impermeable sheeting and leave for 24 hours
5 Immerse the bricks in water for the mortar to harden. For standard tests, the period allowed is such that mortar cubes sampled from that used to fill the frogs and stored in water attain a compressive strength of 4000–6000 lbf/in² (281–422 kgf/cm²) or 28-42 N/mm². This should normally be reached within 3–7 days using ordinary Portland cement, depending on the sand grading

Bricks with frogs in both bed faces Proceed as in Sections 1 to 5, but fill both frogs with mortar of the same composition, allowing an interval (4 to 8 hours for standard tests)

Method (for frogged and unfrogged bricks)
1 Remove the bricks from water and allow them to drain off surplus moisture

2 Check and record the dimensions of the first brick (length and width)

3 Locate the brick centrally on the platen of the testing machine, with a plywood sheet covering each bed face (Fig 38)

4 Apply the load at a uniform rate of 2000 lbf/in² per minute (140 kgf/cm²) or 15 N/mm² and note the maximum load reached

5 Repeat the procedure of Sections 2 to 4 for the remaining bricks.

Results Record these in a table, giving also the calculated *failing stress* of each brick and the average result. Compare the results with Table 25, page 316.

Example 23 Calculate the compressive strength of a specimen to which the following test results apply [a] *Imperial units*:length 8·75 in, width 4·2 in, failing load 52·5 tonf. [b] *Metric technical units:* length 222 mm, width 107 mm, failing load 53·3 tonnef. [c] *Metric S.I.* (*international*) *units:* length 222 mm, width 107 mm, failing load 523 kN (kilonewtons).

[a] Imperial units test area $\quad = 8·75 \times 4·20 = 36·75$ in²
$$\text{failing stress} = \text{load} \div \text{area}$$

$$= \frac{52·5}{36·75} \times 2240$$

$$= 3200 \text{ lbf/in}^2$$

[b] Metric technical units test area $\quad = 22·2 \times 10·7 = 237·5$ cm²

$$\text{failing stress} = \frac{53\ 300 \text{ kgf}}{237·5 \text{ cm}^2} \quad \text{(1 tonnef} = 1000 \text{ kgf)}$$

$$= 225 \text{ kgf/cm}^2$$
$$\text{or } 2·25 \text{ kgf/mm}^2$$

[c] S.I. units: test area $\quad = 237·5$ cm² (as in [b])

$$\text{failing stress} = \frac{523 \text{ kN}}{(237·5 \times 100) \text{ mm}^2}$$

$$= 22 \text{ N/mm}^2$$
$$(\text{or } 220 \text{ } bar)$$

CALCIUM SILICATE (SANDLIME AND FLINTLIME BRICKS)

The raw materials for sandlime bricks, as their name suggests, are *sand*, or *gravel* $\frac{3}{8}$ in (or 9·5 mm) down and *lime*. *Crushed flint* is also used. The lime (calcium hydroxide) is non-hydraulic, but when this is heated by pressurised steam in the presence of sand (silica) it undergoes a chemical reaction to form hydrated calcium silicate, which acts as a binder. For this action to occur the grains of aggregate must be in close contact with the lime, and this is achieved by thoroughly mixing and mechanically pressing the moistened mixture to the shape of the finished brick. The proportion of lime needed varies with the grading of the aggregate and the strength required for the finished brick, but it is usually from 5 to 10% by weight.

Manufacture

The first operation is usually the mixing of the aggregate and lime. If quicklime is used, this is then slaked by the addition of water. If hydrated lime is used, only sufficient water is added to give a mix suitable for pressing, and the raw bricks are sufficiently firm to be handled immediately. The bricks are then conveyed direct to the *autoclave* (steam chamber), in which they remain for a number of hours, depending on the steam pressure and temperature. After leaving the autoclave, the bricks are ready for use. They are made with or without frogs.

Properties

The natural *colour* of the bricks varies with the raw materials, but they are usually off-white, grey or pink. Other colours are obtained artificially by the addition of pigments.

    Calcium silicate bricks unlike clay bricks do not undergo *shrinkage* during manufacture; they are therefore very regular in shape and size. Their natural lightness of colour makes them particularly useful where good light reflection is required. Whereas the method of manufacture usually ensures a consistent quality of production, bricks from different sources can show a wide difference in their properties. The crushing strength of calcium silicate bricks has been shown to relate to their *durability*, so that bricks of high strength are most suited for work to be subjected to a high degree of exposure. Bricks are classified according to strength in BS 187 'Calcium silicate (sandlime and flintlime) bricks'—see Table 28, page 317.

Tests

A summary of BS tests follows.

1 *Dimensions* Bricks must conform individually to given dimensions and tolerances (see Table 29, page 318)

2 *Crushing strength* The test is similar to the test for clay building bricks, but frogs are not filled with mortar and results are based on the minimum net area sustaining the load:

$$\text{Crushing strength} = \frac{\text{failing load}}{\text{gross brick area} - \text{maximum frog area}}$$

Bricks are tested wet (after 18 ± 2 hours immersion) between plywood sheets. BS test limits are given in Table 28, page 317.

3 *Drying shrinkage* The principles of the measurement and calculation of drying shrinkage are dealt with in Volume 1. A typical apparatus used for tests on bricks and blocks is shown in Fig 40.

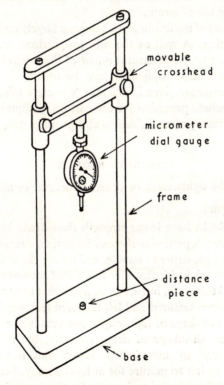

Fig 40 Drying shrinkage test apparatus for bricks and blocks

Metal reference pieces ($\frac{1}{4}$ in/6·35 mm diameter steel ball-bearings for BS tests) are cemented into the two ends of the specimen. The initial measurement is on the saturated specimen, which is then dried at 50–65°C in an oven and re-measured at intervals, after cooling to room temperature in air. BS limits are given in Table 28, page 317.

CONCRETE BRICKS

Concrete bricks are made from a mixture of cement and aggregate (e.g. sand or gravel) together with the least amount of water needed to allow their consolidation by pressing. The low water content enables the bricks to be handled immediately after pressing, and minimises the drying shrinkage which will take place later. A period of air curing allows them to harden before use and this is sometimes reduced by the use of steam.

The properties of the finished brick depend largely on the aggregate type and grading, as well as the mix proportions, water content, and curing efficiency. The natural *colour* of the bricks is determined by the raw materials, but pigments may be added.

BS requirements are given in BS 1180 'Concrete bricks and fixing bricks', and include provision for dimensions, compressive strength and drying shrinkage (see Tables 30 and 31, page 318, for limits).

CONCRETE BLOCKS

These may be lightweight or dense, structural or non-structural.

General properties

Lightweight blocks have lower strength than dense blocks but are superior in such properties as *thermal insulation, ease of cutting* and *nailability*. They are easier to handle, and reduce the deadweight of a structure. Lightweight blocks offer less *sound insulation* than dense blocks of equal thickness, but are more absorbent to sound, provided the surface is open-textured and left unsealed (e.g. unplastered).

The drying shrinkage of lightweight concrete is normally greater than the drying shrinkage of dense concrete, but it can be substantially reduced by autoclaving. Where this is not done the blocks should be left to mature for at least 28 days before they are built in, and preferably moist-air cured for the first 7 days. This will reduce the risk of subsequent cracking. It is also important not to build in blocks which have become substantially wet or saturated, for instance, by rain, as they will inevitably undergo shrinkage on drying out.

CLAY ROOFING TILES

Manufacture

The stages in the manufacture of clay roofing tiles are the digging of the clay, weathering (if necessary), pugging (mixing), moulding, drying, firing and sorting into *firsts*, *seconds* or *thirds*, which means taking account of *underburnt*, *overburnt* and *distorted* products. The drying and firing stages together may take about three or four weeks; longer in the case of hand-made tiles, because of the greater water content of the clay needed for the required plasticity.

Hand-made tiles are formed, face down, by pressing a plastic *clot* into a wood or metal mould, usually onto a sprinkling of sand, which adheres to the tile and is burnt with it. The size of the freshly moulded or *green* tile must allow for shrinkage to the finished size, which, for plain tiles (Fig 41), is typically $10\frac{1}{2} \times 6\frac{1}{2}$ in (265 $\times$ 165 mm). Plain tiles are cambered, usually during the drying stage, by being pressed against a former.

Machine-made tiles are produced using press and extrusion methods. One method for *pressed tiles* is to cut individual tile blanks from a ribbon of clay issuing from a roller. The blanks pass to the press, which forms the tile, together with the nibs and nail holes for hanging. Plain tiles are later cambered before firing.

In the *extrusion process* a ribbon of vacuumised clay is extruded from a die which has the shape of the tile side. In the case of plain tiles, the extruded ribbon includes a continuous nib (Fig 41b). The ribbon is wire-cut to tile width, allowing for shrinkage. Plain tiles are later cambered.

Types of clay roofing tiles

Tiles are made in two main types, known as *double-lap* or plain tiles (Fig 41), and *single-lap*, or interlocking tiles, such as Pantiles and Roman varieties (Fig 42). Tiles which have a flat surface are also referred to as *slates*.

Plain tiles must be laid to partly overlap the two courses immediately below, in order to cover both the nail holes and the side

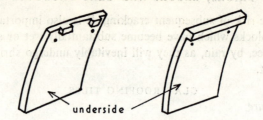

underside

(a) Hand or machine-made, nibbed

(b) Machine-made, extruded, continuous-nib tile

Fig 41 Single lap or plain tile

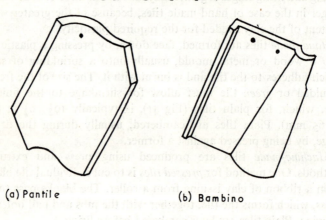

(a) Pantile

(b) Bambino

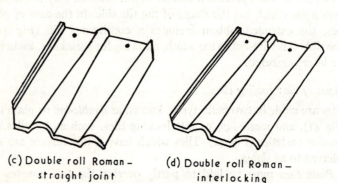

(c) Double roll Roman – straight joint

(d) Double roll Roman – interlocking

Fig 42 Interlocking tiles

joints, at which they are simply butted together. The amount of this lap is commonly 2½ in (64 mm), ensured by fixing the battens from which the tiles (10½×6½ in/265×165 mm) are hung 4 in (100 mm) apart.

Interlocking tiles are specially made to give a side-lap of 1 in (25 mm) or more, and have their edges grooved or ribbed to interlock and exclude rainwater. The headlap is usually 3 in (75 mm), and some tiles interlock with the overlaid course at the top edge. The size of single-lap tiles varies but is considerably greater than that of plain tiles. Advantages claimed for the single-lap tile include the lower weight of roofing, economy of material, wider spacing of tile battens and the option of using a lower roof pitch.

Both single-lap and double-lap tiles are nominally ½ in (15 mm) thick, or minimum ⅜ in (10 mm).

British Standards

Provisions for clay roofing tiles are dealt with in BS 402 'Clay plain roofing tiles and fittings'.

General clauses in BS 402 deal with freedom from *defects*, provision of *nibs*, and *nail holes*, *size*, *thickness* and *camber*. Tiles must be free of unslaked lime and fire cracks, dense, tough, true in

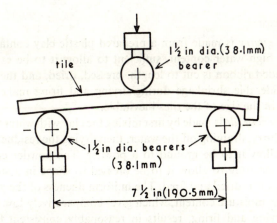

Fig 43 Transverse strength test on a tile

shape, show a clean fracture when broken and be well burnt through-out. Specific requirements apply for transverse strength and water absorption.

The arrangement for a *transverse test* is shown in Fig 43, but details of the apparatus needed are given in BS 402. Tiles are tested wet after 24 hours immersion, with a uniform loading rate of 100–150 lbf/min (45–68 kgf/min, or 445–665 N/min).

For the *water-absorption test,* tiles are first dried at 110–115°C, then cooled, weighed, immersed in water 24 hours, then re-weighed. Results are calculated as the percentage absorption by oven-dry weight.

The BS limits for these tests are given in Table 32 on page 319.

Concrete roofing tiles

These are dealt with in Chapter 8.

CLAY FLOOR TILES AND QUARRIES

Floor quarries and clay floor tiles are both made from the same raw material, but have different characteristics and uses. The tile is thinner and more regular in shape and size. It is also harder and of finer texture and smoother finish than the quarry, resulting mainly from a difference in the method of manufacture.

Manufacture

The *floor quarry* is made from a prepared plastic clay containing a relatively high water content, sufficient to allow it to be extruded. The extruded ribbon is cut to length, pressed, dried, and then fired. The considerable shrinkage during drying and firing makes some variation in the size of the tiles inevitable.

The *clay floor tile* is made by first mixing the clay with water to form a *slip* or slurry, draining off the water, then drying the residue suffici-ently to allow it to be ground to a powder. The powder contains sufficient moisture to allow it to be pressed to shape in a steel die, after which it is dried and fired. The uniform fineness of the powder, and its low moisture content, which gives comparatively low shrink-age in drying and firing, results in reasonably consistent finished dimensions and smooth, regular surfaces.

British Standard requirements

BS 1286 'Clay tiles for flooring designates floor quarries as Class A and floor tiles as Class B.

General clauses deal with the *colour, quality* and *finish* of tiles and a test for water absorption is given. Different *sizes* of tile in each class are specified, ranging up to $9 \times 9 \times 1\frac{1}{4}$ in (32 mm), depending on size, but for all sizes of floor tiles it is 9×9 in (229×229 mm) for quarries and 6×6 in (152×152 mm) for floor tiles. The *thickness* of quarries is from $\frac{5}{8}$ in (16 mm) to $\frac{1}{2}$ in (12 mm). The BS specifies tolerances on dimensions other than thickness, and the tolerances are greater for quarries than for floor tiles.

Use

Both clay tiles and quarries should provide a wear-resistant floor which is easily washed down, but they can be very slippery when wet and because of this are usually used indoors only. Quarries are popular for industrial floors, especially with heavy traffic. They are acid- and oil-resistant, but an appropriate jointing material is necessary, e.g. bitumen or aluminous cement mortar. The joints for quarries are necessarily thicker than for tiles because of the greater dimensional variations.

GLAZED CERAMIC WALL TILES

These are made from specially selected, blended clays and stones. Where china clay and ball clays are used, together with calcined flint and limestone, the finished tile has a white body. Fireclays produce a buff colour.

These materials, after grinding where necessary, are blended with water to form a *slip* of creamy consistency, which is screened to remove coarse particles, then filter-pressed to reduce the water content. The resulting plastic mixture, known as *potters' clay*, is near-dried then finely ground. The tiles are formed by pressing the moist dust to the required shape, making due allowance for subsequent shrinkage.

After preliminary drying, the tiles are fired in a kiln at over 1000°C to form the porous *biscuit*, which is kept until required for glazing. The glaze, which may be white or coloured, is applied by dipping, brushing or spraying. It dries at once by absorption to form a powdery coating, which is fused by a second (*glost*) firing.

Alternatively, the tiles may be once-fired, applying the glaze to the *green* (unfired) tile.

BS 1281, 'Specification for glazed ceramic tiles and tile fittings for internal walls' deals with dimensions and tolerances, and trueness of shape. Tests are given for water absorption and resistance to crazing, chemicals and impact.

SANITARY WARE

The use of clay in the making of such products as W.C. pans, lavatory basins, stall and slab urinals, cisterns and baths is a triumph for technology and workmanship in solving the problems created by the material's considerable firing shrinkage. The larger the piece, and the more complex its shape, the greater is the difficulty in avoiding distortion or cracking during drying and firing. Success has resulted from a combination of improved methods in manufacture and skilful selection and blending of the raw materials. In the process, a number of different types of *ware* have been developed, each with particular suitabilities.

Fireclay

Fireclay is a heavy form of ware, made to withstand the severest treatment in use. It is made from refractory clays of relatively low firing shrinkage, thereby suited to the making of the very largest and heavier pieces, although a long drying and firing period is needed for this.

The raw clay is plasticised with water, then pressed or moulded into shape. Edges are trimmed and component parts butt-jointed if necessary. Before firing, the *green* clay, which is the *body*, is given a coating of *glaze*, applied by brush or spray, to make it impervious; the fired clay itself is porous. The glaze is a specially compounded liquid which, when fired, forms a vitreous transparent skin sealing the buff-coloured biscuit. If white, or coloured, fireclay is required, a special clay mixture, or *slip*, is applied before the glaze.

Earthenware

This is an attractive white, or coloured, ware, of fine-textured, porous material, but with a smooth, highly glazed finish. It will not withstand very rough usage, except for a specially thick grade, known as *heavy earthenware*.

The raw material is a mixture of tempered clay and ground stones with calcined flint, which together form the *slip*. The slip is run into plaster moulds which absorb water and cause the mix to stiffen. After demoulding, final shaping and holing to receive fittings, the article is dipped in a slip of *enamel* (white or coloured glaze), then fired. Firing was formerly done in two stages; the dipping followed the first, or *biscuit*, firing.

Caneware

This is made in a similar way to earthenware, but using ordinary brown clay or marl, which results in a brown- or cane-coloured material. The glaze applied is transparent, but may be preceded by a coat of white or coloured slip on exposed surfaces. Caneware is specially suitable for W.C. pans which have to resist rough usage and exposure to frost.

Vitreous china

This is a denser, stronger material than earthenware. It is vitrified throughout and has a glaze applied on all exposed surfaces. The manufacture is similar to that of earthenware, but china clay (the purest natural clay, which is white and known as kaolin) is used, together with selected stones and calcined flints. Also, firing is at a higher temperature. The greater strength of vitreous china allows it to be used in smaller thicknesses than earthenware, to which it is considered superior in both properties and appearance.

TERRA-COTTA AND FAIENCE

These terms are applied to certain products, such as facing slabs, and blocks for cornices and copings manufactured from refractory clays fired at high temperature. Blocks are usually perforated, made from thin facings and webs to avoid warping and cracking in firing. The two materials are similar but faience is more highly vitrified with a more highly glazed surface.

VITRIFIED CLAY PIPES

These are made from specially blended *plastic* and *ball* (blue) clays; a high firing temperature causes vitrification throughout the material. Pipes are available for surface water and soil drains, and sewers, in various lengths and diameters. They have integral sockets, grooved to receive a mortar joint, or alternatively are provided with rubber rings or plastic mouldings to give a watertight joint with only a push fit. Unless pipes are vitrified throughout, it is normal to require the inside surfaces of these pipes to be glazed. This was formerly done by throwing salt into the kiln, which volatilised to form a glass skin on the pipe surfaces; the product was called *stoneware*. More recently, ceramic glazes have been used. A range of fittings for these pipes is available in the same material.

When referring to clay pipes it is necessary to distinguish between vitrified clay pipes and the shorter *porous earthenware* pipes laid unjointed for *land drainage*.

Summary

Clay products are formed by firing a clay or clay mixture in a kiln, after moulding it to a suitable shape allowing for shrinkage during firing.

Different types of clay are suited to the making of different products, especially with regard to degree of vitrification possible, density and strength obtainable and the amount of shrinkage and distortion likely.

Plastic clays, with a high alumina content, are not suitable for highly vitrified products.

Refractories are made from clays with high silica content. Under-fired clay products generally are unsuitable for severe exposure.

The moisture movement of clay products is very small, but they undergo slight expansion on taking up moisture from the atmosphere after leaving the kiln. They should therefore be allowed to mature at least ten days before use.

Clay bricks and blocks

Bricks may be classified as engineering, facing, common and special types. Engineering bricks have high strength and low water absorption; these two factors are consistent with good durability.

Bricks and blocks of medium or low strength and high water absorption can nevertheless have good durability.

Tests given in BS 3921 for clay bricks and blocks include dimensions, strength, water absorption, soluble salts content and liability to efflorescence. Not all tests apply to every type of unit, and the BS defines units of *special*, *ordinary* and *internal* qualities

Hollow clay blocks are used as walling units and as filler units in in-situ structural concrete floors and roofs. They may be grooved on one or more sides for plaster.

Calcium silicate (sandlime and flintlime) bricks

These are made by autoclaving the raw bricks formed by pressing a mixture of lime and siliceous aggregate. They may be naturally white, grey or pink, but can be pigmented. They have smooth surfaces and are regular in shape and size. The strength of sandlime bricks is an indication of their likely durability. BS 187 classifies sandlime bricks on the basis of strength and drying shrinkage.

Concrete bricks

These are available plain or coloured. Tests are given in BS 1180 for dimensions, compressive strength and drying shrinkage.

Precast concrete blocks

These may be lightweight or dense, structural or non-structural. It is important that blocks not autoclaved should be adequately matured before they are built into work, and are not saturated when used.

Clay tiles

Clay roofing tiles may be hand or machine-made, and include double-lap (plain tiles) and single-lap (interlocking) varieties.

BS 402 gives requirements for clay plain roofing tiles, including tests for transverse strength and water absorption.

Clay floor tiles are thinner, of finer texture and more regular in shape than clay floor quarries. Both should provide a hard-wearing floor surface, but quarries are more suited to heavy, industrial floors.

BS 1286 gives requirements, including dimensions and water absorption, for floor tiles and floor quarries.

Glazed ceramic wall tiles are made from specially blended clay and stone mixtures, either twice-fired (biscuit and glost firings) or once-fired after applying the glaze to the green body. BS 1281 gives general requirements, including dimensions and methods of test (water absorption and resistance to crazing, chemicals and impact).

Sanitary ware

This is made from various types of clay and clay mixtures, according to their thickness and the size, nature and requirements of the product.

Types include fireclay (for large pieces and heavy ware, from refractory clays giving a porous body), earthenware (an attractive ware for general use, from specially prepared clay and stone mixtures, giving a porous body), caneware (for heavy, frost resistant products of limited size from brown clay or marls) and vitreous china (for very strong ware of lustrous finish, prepared from china clay and selected stones giving a vitreous body).

Terra-cotta and faience are made from refractory clays, faience being the more highly glazed.

Clay pipes

These may be vitrified (for surface water, soil drains and sewers) or porous (for land drainage).

10

NATURAL STONE

In recent years the use of natural stone as a facing material, or *cladding*, to buildings has become popular, largely due to the architectural value and beauty of many of our natural stones. In this form of construction, which has to a great extent replaced the use of solid masonry blocks, it is quite common to find the exterior, or interior, face of a building clad with thin stone facing slabs attached to brickwork or other backing, supported by a structural steel or reinforced concrete frame. This is a classic example of a balanced blending of natural and artificial materials to their best advantage.

Fixing stone cladding

These slabs are commonly fixed by metal cramps, anchors and dowels, let in to grooves in the slab and built or cast into the backing, or bolted to the frame of the building. It is usual to leave a clearance of about 10 mm or more between the back surface of the facing slabs and the backing material. With thin slabs, up to about 50 mm thick, this gap is usually left unfilled, and by preventing capillarity over much of the area the weather resistance of an external wall is thus improved. It also reduces the likelihood of salts passing from the backing into the stone facing to cause efflorescence. A further precaution against this is to keep the work dry during the period of construction by covering it if necessary. If *joints* between slabs are left open this prevents water from being trapped behind the facings, otherwise special provision is required to deal with this. These thin facings are likely to undergo high differential thermal movements so that where jointing mortars are used these should not be too strong, and the inclusion of mastic expansion joints is necessary. With thick facing slabs the clearance space is often filled solid with a grout.

More recent methods of fixing stone cladding include the use of synthetic resin adhesives.

Sealing and water-repellent treatments

Sealing the back of stone facings by the application of bitumen, or other paints, either to improve weather resistance or to hold back soluble salts in the backing, can only be successful if done very thoroughly to provide a continuous membrane since any penetration is likely to prove more serious than if no treatment had been given.

The application of water-repellents such as silicones to outer surfaces can give improved weather resistance and durability in some cases, but the effective life of these treatments is uncertain, so that re-treatment may become necessary periodically. Again, such treatments must be thorough.

Geological classification

The geological classification of natural stone may be conveniently considered within three groups, according to the manner of its formation; namely *igneous*, *sedimentary* and *metamorphic*.

IGNEOUS ROCKS

Formation

The igneous rocks were formed by the original cooling of the Earth's crust from its molten state (*magma*), and are also known as *primary* rocks. Their typical structure is of crystalline granular particles of a size and type depending mainly on the composition of the original melt and the rate of cooling. The latter would be largely governed by the depth of the strata beneath the Earth's surface, with the slower rates of coolings at great depths producing larger crystals or grain size, giving a correspondingly coarser texture. This variation leads to a subdivision of igneous rocks into three types, *plutonic*, *hypabyssal* and *volcanic*.

Types of igneous rock

Plutonic rocks are those formed at considerable depth and are consequently coarse-grained. *Hypabyssal* rocks are those formed at intermediate depth, often by the cooling of volcanic magma which had flowed into fissures in the Earth's crust. They are typically of fine crystalline texture. *Volcanic* rocks are those formed at the surface of the Earth, by the rapid cooling of volcanic magma giving a very fine crystalline, or glassy, texture.

Igneous rocks are also classified according to silica content. Those containing more than 66% are termed *acid* and those with less than 52% *basic*. The remainder are termed *intermediate*.

Granites

It is common commercial practice to call any igneous rock which is used in building, a *granite*, and the term is even applied to some of the harder limestones, which are non-igneous. But strictly speaking, all true granites are plutonic rocks of acid character.

True granites are distinguished from other igneous rocks by the fact that they are coarser grained (large crystals are visible without magnification) and generally of lighter colour (e.g. light grey).

Two examples of the many commercial 'granites' available are the *syenites* which are intermediate plutonic rocks of similar texture to true granites, but of a finer grain and darker in colour, and *basalts*, which are basic volcanic rocks of very fine, or glassy, texture, and almost black.

In general, *granites* are the strongest and most durable of all building stones, with very low porosity and water absorption, but they are the most difficult to work. They form the predominant strata in the mountainous regions of the British Isles. Differences in the grain size and mineral content (see below) of stones from different quarries allows a wide choice of colour and texture. In common with all natural stones, they are usually given the geographical name of the locality in which they are quarried.

In addition to their value in heavy constructional and civil engineering work, many colourful granites are highly polished for use as a facing material, where they offer a not-inferior alternative to many polished marbles.

Minerals

Igneous rocks are composed of naturally occurring compounds called *minerals*, and any one rock will contain crystals of one or more different minerals, each of which contributes characteristic properties of colour, hardness and durability.

The chief minerals found in igneous rocks are *quartz* (a crystalline form of silica), *feldspar* and *mica*. (Both of these contain silicates of various metallic elements). True granites contain all three.

The *quartz* is usually distinguishable as glass-like particles which are very hard and cannot be scratched by the point of a penknife or file. Quartz is the most durable constituent of granite.

Feldspar may be white, grey, pink or red. It is quite hard and can just be scratched with the edge of a file or perhaps a sharp penknife. Some feldspars, due to their particular composition, weather less favourably than others, and undergo very gradual decomposition to china clay (kaolin).

Mica varies in colour and may be one of two types, called respectively *muscovite* (white, brown or light green) and *biotite* (dark green to black). Mica is soft and can be cut into with a penknife. It is a flaky material, and when present in large quantities or in large crystals can impair the durability of the stone. The muscovite variety generally gives better durability than the biotite.

Another factor significant to the durability of granites is the widely different thermal expansion coefficients of its constituent minerals, which can lead to disruption under extremes of temperature.

SEDIMENTARY ROCKS

Formation

Sedimentary rocks are also known as *secondary* rocks. They were formed after the cooling and solidification of the Earth's crust and their formation was by one of three agencies; mechanical, chemical and organic.

The *mechanical* formation of rocks resulted from the decomposition of the primary rocks by weathering and erosion. The detritus so produced was deposited over long periods to form successive layers which subsequently compacted to form a solid mass.

The *chemical* formation of rocks may have occurred by the precipitation of calcium carbonate (limestone) from water containing calcium bicarbonate in solution. It is this action which produces *stalactytes* and *stalagmites* often found in caves in limestone regions.

The *organic* formation of rocks occurred as a result of the deposition of countless layers of minute sea organisms, plants or crustaceans. They can be of limestone or silica. Sedimentary rocks include sandstones and limestones.

Sandstones

These consist mainly of *silica*, usually in the form of grains of sand (chiefly of quartz) derived from the disintegration of the primary rocks. The individual grains are themselves virtually indestructible, and the durability of a sandstone therefore depends largely on that of the material which binds the grains together, which may be either *siliceous* (of silica), *calcareous* (limestone), *ferruginous* (of iron oxides) or *argillaceous* (derived from clay). Those with a hard siliceous binder, known as *quartzite*, furnish the most durable sandstones. Of the calcareous types, the most durable are those with the binder in crystalline form, although the durability of all calcareous sandstones is reduced in acid atmospheres. A sandstone which has coarse grains of pronounced angularity is classed as a *gritstone*. Sandstones containing substantial iron oxides are of variable quality, although some are known to give satisfactory service. Small amounts of iron oxides present can greatly influence the colour of the stone. Reds, yellows and browns are common. In general, argillaceous sandstones are not sufficiently durable to be used for building work.

Limestones

These are of calcium carbonate (*calcite*), magnesium carbonate (*magnesite*) or both (*dolomite*), usually together with 'impurities' such as iron oxides, sand or clay. Different limestones show wide variation in durability, and whereas the softer, more porous varieties (e.g. chalk) are usually unsuitable for most building work, some of the harder, more crystalline types (e.g. some *carboniferous* limestones) are similar in texture, density and strength to igneous rocks or marbles. The colour of limestones varies from white, grey, blue and pink to yellow.

A simple test sometimes used to assist identification of limestones is to add a little dilute hydrochloric acid and watch for effervescence. However, since a reaction will also occur with a *calcareous* sandstone, the test is more useful in isolating siliceous sandstones. This action is a reminder that limestones may weather badly in acidic industrial atmospheres.

Use of sandstones and limestones

Many varieties of sandstone and limestone suitable for building work are quarried in the British Isles, and many of these have considerable architectural value.

A feature of the sedimentary stones is the stratification which has occurred in their formation. This is usually visible by the layers of different colour or texture which formed the original *bedding planes*. These stones often have *planes of cleavage* along which they easily split, but it should be noted that these are often not parallel to the original bedding planes, because earth movements have changed the original direction of pressure.

As a general rule, limestone and sandstone should not be used together in the same construction because of the destructive chemical action which can take place. Rainwater which has become slightly acidic from dissolved sulphur gases slowly dissolves calcium carbonate to form water-soluble calcium sulphate. The solution of calcium sulphate is drawn into the adjacent sandstone by its greater capillary pull; the pores of sandstone are usually finer than the pores of limestone. On drying out, this solution deposits crystals near the surface of the sandstone, which ultimately causes exfoliation.

METAMORPHIC ROCKS

The name *metamorphic* is given to rocks which have undergone a change from their original form by the action of high pressure or high temperature, or both. The two important examples are *marble* and *slate*.

Marble

True marble is a crystalline, metamorphic form of limestone. Commercially, however, most limestones which are sufficiently hard to take a polish are classified as marbles.

The presence of coloured veins in marbles is due to impurities, such as iron oxides, and can result in a highly attractive appearance.

Slates

Slates are metamorphic forms of clays and shales formed from the detritus of weathered igneous rocks; or they can also consist of metamorphosed volcanic dust.

Slates have a laminated structure and are easily split into thin slabs which have been widely used in the past as a roof covering and as a damp-proof course because they are highly impervious. Their use for roof covering and damp-proof courses has recently greatly diminished for economic reasons and the availability of alternative materials. In addition to this, slate does not allow the degree of flexibility often required in a d.p.c. Slate is now widely used as cladding, and will take a polish. The finer grained varieties give the flattest cleaved surface, and consequently can be split to the thinnest sheets.

There are a number of sources of slate in Great Britain, mainly in the western regions, and they include both open quarries and mines.

A good quality slate should not be delaminated by frost action. Some are not durable in acidic atmospheres, and they should be selected accordingly.

TESTS FOR STONE

The *durability* of a building material depends upon its capacity for endurance, and this is particularly important with building stones. The results of laboratory tests are not necessarily conclusive and at best can give only limited guidance on this point. Stone should therefore be selected on the basis of experience or knowledge of previous performance in work of a similar type in similar conditions.

Tests which can be applied to stone include the following:

1 *Compressive strength* (usually of cubes 150 mm, 100 mm or smaller size, cut to shape, or of cylinders—cut or drilled)

2 *Porosity, water-absorption, permeability and specific gravity,* using the normal methods of test, e.g. as for bricks and tiles

3 *Frost resistance.* The usual test is to subject saturated specimens to alternate freezing and thawing (10 or more daily cycles), observing any effects such as delamination, loss in weight or loss of strength

4 *Acid resistance.* This test is only significant where the stone is to be used in acidic atmospheres. A standard test for slates is to immerse 50 mm square specimens in dilute sulphuric acid (1 volume of con-

centrated acid, s.g. 1·84, to 7 volumes of distilled water). Of 3 specimens, none should show delamination, swelling, softening or flaking, nor evolve gas during immersion for 10 days

5 *Crystallisation test*. This test aims to reproduce the conditions associated with exfoliation due to calcium sulphate. However, since calcium sulphate is only slightly soluble and the test is an accelerated process, sodium sulphate is used (14% by weight solution). Cubes, size 40 mm, are dried, weighed and immersed for 2 hours then oven-dried overnight. After 15 cycles the percentage loss in weight is determined

6 *Chemical tests*. These can assist geologists in identifying minerals and other compounds present

7 *Microscopic examination*. Sections mounted on a glass slide are reduced to a thickness at which they are virtually transparent (about 0·02 mm). Immersion in a staining solution causes the pores to fill and become visible so that their size and distribution can be assessed

Summary

Stone is widely used in building as a cladding material and also as masonry blocks. The three main geological classes are the igneous, sedimentary and metamorphic rocks.

The *igneous*, or *primary* groups, formed by the original cooling of magma, includes the plutonic (deep seated) rocks—e.g. true granites, the hypabyssal rocks—formed at intermediate depth and finely crystalline, and the volcanic rocks, formed at the surface and very finely crystalline or glassy.

Igneous rocks are also classified by silica content as acid (more than 66%)—e.g. true granites, basic (less than 52%)—e.g. basalts, or intermediate—e.g. syenites.

True granites contain the three minerals, quartz (very hard), feldspar (hard) and mica (soft), which have different thermal expansion coefficients and impart different characteristic colours.

Granites are used in civil engineering construction, as masonry blocks, as thin facings for cladding and where their very low permeability is useful.

Sedimentary, or *secondary*, rocks are those formed by the consolidation of sediment produced by mechanical, chemical or organic means, and include the sandstones and limestones. They have bedding planes and planes of cleavage.

Sandstones are composed mainly of quartz (silica) grains and their durability depends mainly on the nature of the cementing agent (siliceous, calcareous, ferruginous or argillaceous).

Limestones may be of calcium carbonate (calcite), magnesium carbonate (magnesite) or both (dolomite). Different types vary considerably in their durability, according to porosity, hardness and the degree of crystallisation.

Sandstones and limestones should not be used together.

Metamorphic rocks are those re-formed under the influence of high pressure or temperature and include the marbles (from limestones) and slates (from shales or volcanic dust).

The *durability* of building stones is usually known from experience of their previous use in buildings.

Tests on stone are for compressive strength, porosity, water absorption, permeability, specific gravity, frost resistance, acid resistance, crystallisation damage and microstructure.

11

TIMBER

Timber is valuable as a building material for a number of reasons. It is structurally useful because of its high strength in relation to its density. It is comparatively easy to work to a variety of shapes either by hand or machine, is durable under appropriate conditions, and can give a good finished appearance at reasonable total cost. However, as it is a product of Nature which comes to us in countless varieties of species and qualities, it must be properly processed and selected to suit the work in hand.

Timber is an *organic* material (i.e. of carbon compounds) produced in the growth processes of a living tree. The main source of timber used in the building industry is the *trunk* (main stem) of the tree. The *girth* (circumference) of the trunk increases as the tree develops, and when the tree is thought to have reached maturity it is felled and the raw timber is *converted* (cut to baulks and planks) and *seasoned* (partially dried) before use.

GROWTH, COMPOSITION AND STRUCTURE

The growth and structure of timber is most clearly illustrated by the three principal trunk sections (Fig 44)—the *cross section* (or transverse section), *radial section* and *tangential section*. The annotations to Fig 44 will now be explained.

Growth

Growth takes place outwards from the centre of the trunk (except in bamboo and palm tree types, with which we are not concerned) and a new outer sheath of *wood tissue* is deposited during each growth period, or *season*. Each sheath, or layer, is termed a *growth*

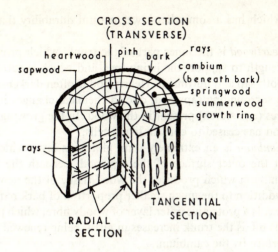

CROSS SECTION
(TRANSVERSE)

heartwood   pith   bark   rays
sapwood   cambium
                           (beneath bark)
                           springwood
                           summerwood
                           growth ring

rays

TANGENTIAL
SECTION

RADIAL
SECTION

Fig 44 The three principal trunk sections

ring, and these concentric rings are in some cases clearly visible on inspection of the cross-section.

In temperate climates, as in the British Isles, growth takes place mainly in the spring and summer months, so that each growth ring normally represents one year's growth, and may then be referred to as an *annual ring*. In some tropical climates, however, the growing season is almost continuous and it is possible to have more than one growth ring during a year. Trees of rapid growth will tend to have wide growth rings, whereas slow growth is associated with narrow growth rings.

Growth in summer is slower than it is in spring, and this results in the *summer wood* (also called *late wood*) being stronger, and usually darker than *spring wood* (*early wood*). This variation in colour, or texture, is one means by which the growth rings are rendered distinguishable.

The *pith*, or heart centre, is a fibrous core of woody tissue forming the core of the trunk, which decays as the living tree gets older, and then has no practical value.

The *sapwood* is the more recently formed wood tissue comprising the outer growth rings. It contains a large amount of sap (aqueous solution) and nutriment (food) for the growing tissue, and furnishes timber which has a somewhat lower natural durability than heart-wood.

The *heartwood* is the inner part of the trunk, which provides the main strength to support the living tree, and which is substantially free of both sap and nutriment. Heartwood is often darker in colour than the surrounding sapwood, due to chemical changes in certain substances (e.g. starch) present in the wood. In the growing tree the heartwood has ceased to contain living cells.

The *cambium* is an extremely thin, glutinous layer which forms a film at the outer surface of the sapwood, beneath the bark. It is the cambium which produces the wood tissue of the new growth ring, in addition to the much smaller proportion of bark required.

The *bark* is a protective outer layer of woody fibre, which periodic-ally scales off as the trunk increases in girth, being renewed from its inside surface by the cambium.

Wood composition

Wood tissue is composed of various types of *cell*, the units of which living matter is composed. These cells have various shapes, but all consist of a *wall* enclosing a *cavity*. In the living tree the cavity will usually contain liquid matter. The living cells of the cambium are filled with a watery liquid called *protoplasm*. Cells produced by the cambium undergo a process known as *lignification* to form wood, in which the cell walls thicken.

There are essentially three functions to be performed by the various types of wood cell. These are to conduct the sap, to provide storage space for food, and to provide the mechanical strength necessary for the support of the tree. To produce the substance of which the cells and foodstuffs are composed, the tree must first of all manufacture its own *organic* matter, the stuff of which all living tissue is made, and it achieves this by rather fascinating means. In the spring, *sap*, which is water containing various mineral salts in solution, is taken up from the soil by the roots of the tree and passes up through the sapwood into the branches. From the branches it passes into the leaves, where it is converted into *carbohydrates*, organic compounds of carbon, hydrogen and oxygen which are the tree's raw materials for foodstuffs, by a process known as *photo-synthesis*. In this process, carbon dioxide is absorbed from the atmosphere and undergoes chemical action with the sap, using the sun's light as energy and the green colouring matter of the leaves (*chlorophyll*) as the catalyst (activator). Food material in solution then passes from the leaves back down the trunk through an inner layer of bark (the *bast*) next to the cambium. Part of the food material will be consumed in the form of *sugars*, some being converted to *cellulose*—the important structural component of wood tissue, and part will be held in reserve as *starch*, for conversion to sugars when required.

Chemical nature of wood

The cell walls of wood are composed of cellulose, which, like starch and sugars, is a carbohydrate, and a substance called *lignin* (the non-carbohydrate constituent). The lignin acts as a cement, binding together the cellulose material in the cell walls to give the wood its strength and rigidity.

Softwoods and hardwoods

These are the two broad classes into which timbers are divided commercially, according to whether they are from *conifers* (pine trees or firs), or from *broadleaf* trees. A botanist would call these 'gymnosperms' and 'angiosperms' respectively.

The conifers furnish *softwood* timbers which have needle-like leaves and are cone-bearing. The cones are exposed seeds. *Hardwoods* are furnished by broadleaved trees which have covered seeds. Their name is descriptive, and distinguishes them from the conifers as previously defined. The two terms can be misleading, since although most hardwoods are in fact mechanically harder than most softwoods, this is not invariably the case. For example, *pitch pine*, although classified as a softwood, is mechanically harder than some hardwoods, whereas *balsa* and *willow* are notable examples of very soft hardwoods. Another distinction is that the hardwoods are mainly *deciduous* (lose their leaves in winter) whereas softwoods are mainly *evergreens*.

Apart from the differences mentioned in defining softwoods and hardwoods, there are important differences in the cellular structure of the two groups.

Softwood structure

The softwoods have the more primitive structure, with only two types of cell, *tracheids* and *parenchyma* cells. Enlarged sections of these are shown at Figs 45 and 46. The *tracheids* are pod-like, form the bulk of the timber, and are longitudinally arranged in the trunk and radially distributed about the pith. They have the dual rôle of conducting the sap and providing the tree's mechanical strength, for which purpose the cells develop thicker walls. *Parenchyma cells* provide storage space for foodstuffs. They are brick-shaped and butt against one another to form chains, which are mainly horizontal and lie radially about the pith and are therefore called *rays*. However, they may also be in vertical chains, known as *wood*, or *strand*, *parenchyma*. Tracheids sometimes occur lying horizontally, in association with rays, and are then called *ray tracheids*.

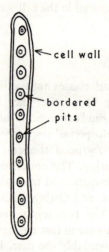

Fig 45 A tracheid (section)

Conduction between the different cells takes place through *pits*, which are areas of the cell walls which, because they have not undergone thickening during the process of lignification, are permeable to fluids. The two basic forms of pitting are the *simple* pit and the *bordered* pit. Fig 47 shows these as a section through the cell wall. They are seen in elevation at smaller scale in Figs 45 and 46.

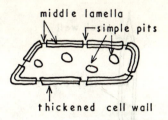

Fig 46 A parenchyma cell (section)

The simple pit has different shapes according to the species of timber in which it occurs and it is these features which allow the positive identification of a particular species under microscopic examination. Bordered pits, which occur only in tracheids, can act as a valve controlling the flow of sap or foodstuffs by movement of the disc-shaped thickened central region (the *torus*) in response to pressure.

Hardwood structure

There are four types of cell in hardwoods and they include fibres and vessels, in addition to parenchyma and tracheids.

The *fibres*, which are needle-like in form (Fig 48) make up the bulk of the timber, and act as mechanical tissue.

The *vessels* are cylindrical tubes (Fig 49) forming continuous vertical chains in the trunk to provide for sap conduction. The

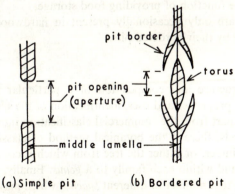

Fig 47 Section through a cell wall at pits

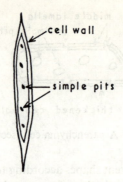

cell wall

simple pits

Fig 48 A fibre (section)

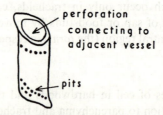

perforation connecting to adjacent vessel

pits

Fig. 49 A vessel

circular or oval section of a vessel as seen at a cross-section is termed a *pore*.

The *parenchyma* cells are of the same type as those in softwoods, with the same function of providing food storage.

Tracheids are only occasionally present in hardwoods and are distinguished by their bordered pits.

Identification

Since the structure and composition of a particular timber will determine its properties and uses, identification of the species can be important. Apart from the commercial classification into softwoods and hardwoods, there is the botanical method of classification, in which each timber, or rather the tree from which it comes, belongs to a *family*, and within each family to a *genus*. Finally, each genus may give rise to a number of different *species*, which may be given a common, or local, commercial name.

For example, within the pine family (*Pinaceae*), there are a number of different genera (plural of genus), including those of *Pinus* (the 'true' pines) and *Picea* (the spruces). Within each genus are a number of different species, such as *Pinus sylvestris* (Scots pine), *Pinus strobus* (Canadian yellow pine) and *Pinus palustris* (American long-leaf pitch pine). Similarly, there are *Picea abies* (European spruce or whitewood), *Picea glauca* (Canadian spruce) and *Picea sitchensis* (Sitka spruce).

To take a further example, this time from the hardwoods, in the beech family (*Fagaceae*), are found the genera *Fagus* (the 'true' beeches), *Quercus* (the oaks) and *Castanea* (the chestnuts), as well as the genus *Nothofagus*. Again, there are different species, e.g. *Fagus sylvatica* (European beech), *Fagus grandifolia* (North American beech), *Quercus robur* (a European oak), *Quercus rubra* (American red oak) and *Quercus cerris* (Turkey oak), *Castenea sativa* (European chestnut) and *Castanea dentata* (American chestnut). Examples of species belonging to the Nothofagus genera are *Nothofagus dombeyi* (coigue beech) and *Nothofagus procera* (rauli).

The botanical method of classification is useful because it is used throughout the world, and is of practical value in that timbers within the same family will have certain common features, and those within the same genera may have similar or even very similar properties, and therefore, similar uses.

Timber species may sometimes be identified by their general features; for example, by colour, density, smell, the appearance of their growth rings or rays when visible, or the shape and distribution of their pores, if present. However, more detailed examination is often needed, such as by magnification under a hand lens or microscope.

The *hand lens* is most useful for examining hardwoods, and one giving a magnification of ten (×10) is often used. In the case of softwoods, there is usually little to be gained by hand lens examination beyond what is already visible to the naked eye, and for these the much greater magnifications obtainable under a microscope are invaluable.

EXPERIMENT 28 (Fig 50) Examination of hardwoods and softwoods by hand lens

Apparatus Hand lens (×10)

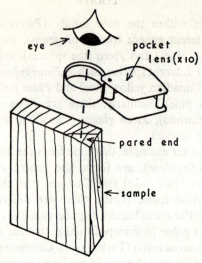

Fig 50 Timber examination by hand lens (Experiment 28)

Specimens Hardwood and softwood samples, size approximately 100 × 50 × 10 mm with the 50 × 10 mm edge showing a cross-section (Fig 50)

Preparation Cut a small area of the end grain of each specimen cleanly by paring it with a very sharp cutting edge such as a chisel or wedge-shaped cutter

Method

1 Hold the lens near one eye and bring the cut end of the specimen towards it until clear focus is obtained. It should not be necessary to close the other eye

2 Examine the section, observing the main features (see *note* below), and sketch, in pencil, what you see. Do not attempt to reproduce exact detail; a likeness is all you need. This is shown in Fig 51, where typical distinguishing features are illustrated

3 Repeat the procedure (Sections 1 and 2) for the remaining specimens

Note A cross-section should always be held and drawn with the growth rings horizontal, when it is being examined and sketched. The rays should be vertical with the top and bottom edges of the

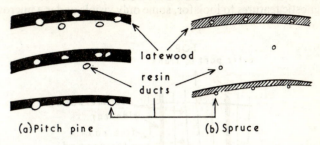

SOFTWOODS

latewood

resin
ducts

(a) Pitch pine   (b) Spruce

HARDWOODS

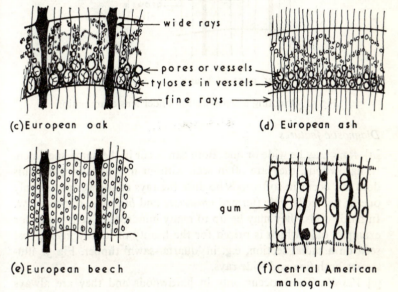

wide rays

pores or vessels
tyloses in vessels
fine rays

(c) European oak   (d) European ash

gum

(e) European beech   (f) Central American
                                       mahogany

Fig 51 Typical cross-sections illustrated

section facing the outer and inner parts of the trunk respectively, as
shown in Fig 52.

Whereas the rays, or other features, may, in fact, be lighter in
colour than the surrounding areas, in a line drawing they are usually
shown as dark lines

Results Recognising and interpreting the main features of a specimen requires some practice and experience, and a summary of the main diagnostic features to look for, some only visible under a microscope, follows.

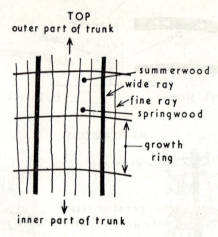

Fig 52 Orienting a cross-section under lens examination

Diagnostic features

[*a*] *Rays* may be wide or fine. Both can occur in the same specimen. In hardwoods, they are often seen without magnification, but this does not occur in softwoods because the rays of softwoods are only one or two cells wide (termed *uniseriate* and *biseriate*, respectively). In hardwoods, they may be up to many hundreds of cells wide; for example, oak, which is prized for the beauty of its ray *figure*, best seen on a radial section, e.g. in 'quarter-sawn' timber. Fig 52 illustrates both fine and wide rays.

[*b*] *Vessels*. These occur only in hardwoods and they are always present. They are seen on a cross-section as circular, oval or angular cavities—the *pores*—varying in size according to the species and the tree's rate of growth. An important distinction is made between what are called ring porous and diffuse porous timbers.

In a *ring porous* timber the pores at the commencement of a growth ring are distinctly larger than those elsewhere and form one or more complete, or nearly complete, chain rings (Fig 53a). Only a few timbers are ring porous; examples of these are *oak*, *ash*, *elm*, *teak*, *sweet chestnut*, *cedar* and *walnut*. In some cases, however,

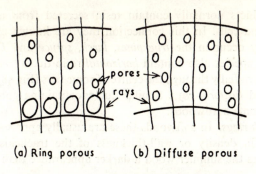

(a) Ring porous   (b) Diffuse porous

Fig 53 Types of pores (in hardwoods only)

the chains of large pores may be substantially incomplete, and the term *semi-ring porous* is then used. This can occur in the case of *teak* and *sweet chestnut* as well as in species of *cedar* and *walnut*, especially in European walnut, *Juglans regia*.

In a *diffuse porous* timber (Fig 53b) the pores undergo no sudden change in their size throughout a growth ring. They may be of uniform size, change gradually in size, or they may be irregularly distributed in different sizes. Examples are *alder, beech, birch, box, willow* and *sycamore*, together with many *tropical hardwoods*.

Pores may also occur either singly or in groups. These are described as *solitary* or in *multiples* respectively.

[c] *Resin ducts.* Vertical resin ducts (Fig 54) are seen as tiny holes on the cross-section of certain *softwoods*. They are usually just visible to the naked eye. They are not easily mistaken for the pores of hardwoods, since under magnification their edges are less sharply defined than in pores, and they occur in isolated groups, or singly, in part of the growth ring only (Fig 54).

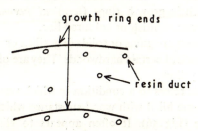

Fig 54 Vertical resin ducts in a cross-section

G

Resin ducts normally contain resin secreted from a lining of parenchyma cells. In this connection they are known as *epithelial* cells. They occur in the *true pines, larch, spruce* and *Douglas fir*. The same species also contain *horizontal* resin ducts (*resin canals*), which run radially through the centre of some rays (Fig 58) which are then called *fusiform rays*. Similar ducts, containing gum or resin, occur in some hardwoods.

[*d*] *Growth rings*. In softwoods these are usually apparent due to a variation in density or wall-thickness of the tracheids (Fig 55), which gives the summer wood a darker appearance than the spring wood.

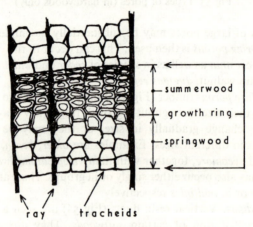

Fig 55 Growth ring in a softwood

In hardwoods, the growth rings may be apparent because of a ring porous structure, a variation in pore size, or a pore-free zone. Also, some hardwoods have concentric bands of wood parenchyma at the end of their growth rings (*terminal parenchyma*). This is frequently seen as a whitish or coloured line.

[*e*] *Deposits*. These are gums or solids (e.g. silica or calcium oxalate) found in the vessels of certain hardwoods. They are often distinctively black, white or yellow.

[*f*] *Tyloses*. This refers to a condition in which vessels in the heartwood have become filled with wood substance which has a sponge-like appearance (Fig 56). It often appears to glisten under lens examination.

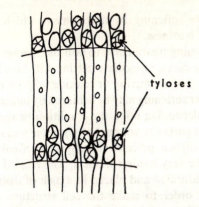

Fig 56 Tyloses in hardwood vessels

Microscopic examination

This is done by cutting extremely thin slices or 'sections' of the wood about 0·01 mm thick and viewing them by passing light through them into a microscope. These sections are very fragile and therefore have to be *mounted*, that is, supported and held intact by being fixed between two thin glass sheets (a *slide* and a *coverslip*) by a jelly, or wax-like substance (e.g. glycerine jelly or Canada balsam). It is usual to mount the three principal sections of a sample (transverse, radial and tangential) on one slide, say 75×25 mm, using one coverslip, say, 40×20 mm (Fig 57). The sections are usually cut from a 10 mm cube of timber. Cutting is most easily done using a special machine called a *microtome* (sledge type), but manual cutting using a flat-ground razor of wedge section, or in some cases even a very sharp chisel, can be satisfactory. Before cutting, the timber will

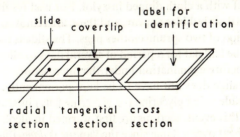

Fig 57 Arrangement of a timber slide

normally require softening by a treatment which varies with its initial degree of hardness.

A simple softening treatment is to boil the specimen in water until it sinks. In some cases, immediate cutting will then be possible, but in others a period of simmering or soaking will be required. Where necessary, further softening may be achieved by immersion for several days or longer depending upon the condition of the specimen, in a mixture of equal parts (by volume) of *glycerine* and *methylated spirit* (industrial alcohol) in a specimen tube. Special softening methods may be needed for the very hardest timbers. The treated specimen is cut while it is still saturated and placed in a dish of distilled water. It is then stained in order to make the cell structure show up under examination. Different stains may be used to bring out different parts of the cell structure, and the method of staining varies. One simple procedure is to immerse the specimen for several minutes or longer in a 1 % aqueous solution of *safranin*, which stains lignin red, in a watch glass or evaporating dish, then, subsequently rinse it thoroughly in distilled water to remove surplus stain. The specimen should then be twice rinsed in 98% alcohol for 2 minutes, once in absolute alcohol (1 minute) to remove water and finally in pure *xylol* (1 minute) to clear it. If the xylol appears clouded, a further rinse in absolute alcohol, followed by one in xylol, should be given. Prepared sections (transverse, tangential and radial) may then be mounted in Canada balsam dissolved in xylol to give a treacle-like consistency (obtainable ready-mixed). This is done by smearing a line of balsam along both the slide and coverslip and laying the cut timber sections on the slide, leaving room for a slide label. One of the shorter edges of the coverslip is then stood on the slide and is lowered like a hinged trap-door. The slide is pressed gently to squeeze out surplus balsam, which can be wiped off with a cloth dipped in xylol. For best results the slide is warmed until air bubbles appear, and these are removed in squeezing by application of two sprung clothes pegs. The slide is then complete, but should be laid flat for a period of several days to allow the balsam to harden before examination under the microscope.

Where a slide does not have to be kept permanently, a *temporary* mount is made using glycerine (or glycerine jelly) instead of Canada balsam. In this event, sections are first warmed in alcohol to remove air (do *not* heat over a flame since the vapour is *dangerously explosive* in air, warm by partly submerging a test-tube of alcohol in hot water),

then well rinsed in distilled water. Glycerine jelly should be warmed
before use. Separate staining can be avoided by mixing a little aqueous
stain (e.g. 1% *Gentian violet*) with the warmed mounting medium.

The typical structure of a softwood, as seen from radial, tangential
and cross-sections viewed separately under a microscope, is shown in
Fig 58. To illustrate the use of such sections, an abridged key to aid
the microscopical identification of softwoods in common use in
Great Britain is given in Table 33 (Parts I and II) on page 320.
This is useful in grasping the principles of identification, but
remember that more detailed information can be obtained from
published works which give a comprehensive treatment of the
subject.

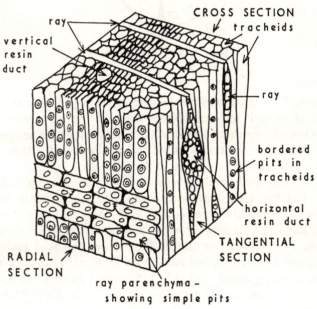

Fig 58 Microscopic structure of a softwood

CONVERSION AND SEASONING

Conversion

Conversion refers to the preliminary cutting and trimming of the
felled trunk into baulks or planks of timber, to allow easier handling

and to prevent the development of serious defects due to unequal drying shrinkage throughout the trunk. If left to dry unconverted, a felled trunk may develop radial cracks known as *star shakes* (Fig 59). The sapwood shrinks more than the heartwood on drying due to its greater initial moisture content in the form of sap, and its faster rate of evaporation.

Fig 59 Star shakes in a trunk

Seasoning

Seasoning is the process of removing moisture (sap) from converted timber to bring it into equilibrium with the relative atmospheric humidity under which the timber will be used. It entails the removal of all the *free moisture* present within the cell cavities, together with a part of the moisture contained in the cell walls.

When felled, the 'green' timber may contain anything up to about 150% of its dry weight of moisture as sap. When all the free moisture is removed, there will still be about 25% of moisture in the cell walls. This is known as the *fibre saturation point* and part of this must be removed to condition the timber for its use. The final moisture content required may be between about 18 to 24% for carcassing timber, down to about 7 to 16% for internal joinery. The lowest values apply to timber for internal use in buildings with a high degree of central heating.

The reason for seasoning as far as possible to the *equilibrium moisture content* is that changes of moisture content below fibre saturation point result in *moisture movement* (shrinkage and expansion), which can give rise to defects such as splitting or warping. Under-seasoned timber will shrink on losing moisture by evaporation. On the other hand, timber dried to a moisture content below that required for equilibrium with the surroundings will take up moisture and swell because of its natural *hygroscopic* tendencies.

In seasoning, timber must be dried gradually and uniformly throughout to avoid setting up stresses due to unequal shrinkage.

Most timber used in building nowadays is *kiln-seasoned*, the kiln being a specially built chamber affording controlled conditions of temperature, humidity and air circulation. This offers the advantages of speed and finely controlled conditions over the older and now less-used method of *air-seasoning*, in which the green timber is stacked under cover in the open with provision for the free circulation of air between the individual pieces. A combination of the two methods is also used where timber is first partially air-seasoned, then kilned to the final moisture content required.

EXPERIMENT 29 Moisture content of timber (Fig 60)

Note The method given here is one used in commercial practice

Apparatus Chemical balance (weighing to 0·01 g)—drying oven (ventilated), maintained at 100–105°C—fine-toothed saw

Specimen A sample plank taken from a batch of timber of which the moisture content is required

Method

1 First cut a length at least 9 in (200 mm) from the end of the plank and then discard it as it may be drier than the bulk of the timber

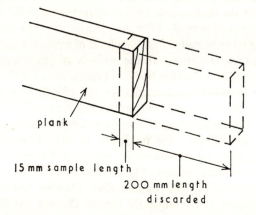

plank

15 mm sample length

200 mm length
discarded

Fig 60 Sampling timber for a moisture content test

2 Cut a strip free of knots 15 mm wide from the same end of the plank and immediately weigh it. This will be the *initial weight*

3 Dry the strip to a constant weight in the oven and note the final weight. This is the *dry weight*

Result Calculate this from the following formula:

$$\text{Moisture content (per cent)} = \frac{(\text{initial weight} - \text{dry weight}) \times 100}{\text{dry weight}}$$

DURABILITY OF TIMBER AND PRESERVATIVES

Decay of timber by fungal attack

The decay of timber is caused principally by the attack of fungal growths (*fungi*), of which there are many different varieties. Some of these attack only the green timber (the living tree, or felled timber), whereas others prefer seasoned timber.

A *fungus* is a plant growth which is unable to produce its own organic material by photosynthesis, since it has no chlorophyll (see page 185). It therefore feeds on decaying or living vegetable or animal matter, in this case timber.

Fungal growths develop by the germination of microscopic *spores* produced in enormous numbers by the fruiting body of an existing growth. These spores correspond to the seeds of a normal green plant. The spores are released to float in the atmosphere, and after settling on timber will remain dormant until the right conditions for growth prevail. After germination, growth occurs in the form of fine strands, called *hyphae* (plural of hypha), which together form a mat or *mycelium*. If the growth of mycelium becomes sufficiently prolific a fruiting body will form.

The four essential requirements for the growth of a fungus are *oxygen* (or air), *moisture*, *food* (timber) and favourable *temperature* conditions. The optimum conditions for moisture content and temperature vary according to the particular variety of fungus, but, in general, damp, warm conditions are preferred. For example, timber which has a moisture content of less than about 20% is known to be immune from attack by the common *dry rot* fungus, and this level of moisture content is consequently termed the *dry rot safety line*. The temperature largely controls the rate of growth of the different fungi, with the optimum range for most of them in the region of 20 to 30°C. Growth is halted at or below freezing point, but only while that condition is maintained. On the other hand, fungi will cease to live if the temperature rises above about 70°C, when the timber is said to be *sterilised*. In practice, this means that all timber is effectively sterilised during kilning, although this does not make the timber immune from a later attack.

Types of fungi

The two main classes of wood-rotting fungi are the 'brown' rots and the 'white' rots. The names refer to the colour of the timber when it is in an advanced state of attack. The white rots may cause slight initial darkening but this is followed ultimately by a permanent whitening of the attacked timber.

Brown rots

Brown rots are cases where the *cellulose* of the cell walls is attacked, but not the lignin. The fungi responsible include the true *dry rot* fungus (Merulius lacrymans) and the *cellar-fungus* (Coniophora cerebella).

The common dry rot is characterised by the appearance of the decayed timber (usually a softwood), which is light brown and shows a cubical pattern of cracks along and across the grain. The attacked wood eventually crumbles to a dry powder. The growth has a characteristic musty smell. Although a damp atmosphere is required to support the growth, persistently wet timber cannot be attacked, and this fungus is often found growing in houses or other buildings. Dry rot is notorious for its ability, once a growth has started, to extend its attack to *dry* timber at some distance from its source of moisture. To do this, its mycelium is able to bridge other materials and can even penetrate brickwork. The mycelium is white or silver grey (but sometimes with a lilac or yellow tinge), with a soft, cotton-wool texture. The fruiting body appears as a large fleshy 'pancake' with a white margin and a rust-red centre.

The *cellar-fungus* can only develop and flourish in saturated or very damp timber, and so it belongs to a class known as *wet rots*. The decayed timber (softwood or hardwood) appears dark brown, with cracking principally along the grain. Attack can develop internally without the surface being visibly affected. The mycelium consists of fine strands, initially yellowish, later becoming brown to black. The fruiting body is a thin, irregular shaped skin, olive green to olive brown, with a pimply appearance.

White rots

White rots are fungi which attack both the *lignin* and the *cellulose* of the cell walls. This class includes many of the fungi which attack felled timbers in the forest, and external fencing.

Surface moulds (*mould fungi*)

Surface moulds may appear as thin, powdery, or cotton-wool like areas on timber surfaces. They are commonly white, blue or green, sometimes with scattered black dots. They are caused by types of fungi which feed only on the cell contents and do not attack the wood substance forming the cell walls. They consequently do no mechanical damage to the timber, but appear unsightly and may cause staining of the surface of the timber. They also indicate that conditions are favourable to attack by one of the wood-rotting fungi. The moulds are usually easily removed by brushing, and any stained timber can be removed by sanding or planing.

Sap-staining fungi

Like the mould fungi, sap staining fungi feed only on the cell contents. They affect only the sapwood, causing it to appear dis-coloured, usually blue, bluish-grey or brown. Infection often occurs in the felled logs while still in the forest, and since the fungus dies during kilning the only real disadvantage is the discolouration, which is usually non-uniform and of 'blotchy' appearance. Active growth of these fungi can only take place when the conditons are appropriate to the development of wood-rotting varieties.

Mere discolouration does not necessarily indicate the presence of a sap staining fungi. For example, *chemical staining* may occur due to the acidic or alkaline nature of some of the minor components of certain timbers. For example, the tannin in such timbers as oak, Douglas fir, sweet chestnut, afrormosia and African mahogany reacts with iron to cause *iron stain*.

Dote

It is sometimes difficult to distinguish the effects of certain stains or surface moulds from those of the early stages of an attack by one of the wood-rotting fungi, known as *dote*. Dote often appears as streaks or patches of brown, red, grey or white discolouration. A simple test is to insert the tip of a knife blade into the surface of the affected timber and lever up a splinter. Wood affected by rot will fracture comparatively easily across the splinter, whereas a splinter from sound timber will tear up the adjacent wood. In some cases of wood rot the discoloured area can easily be indented by the pressure of a thumb-nail.

Preservatives

Timber preservatives offer positive protection against fungal attack by poisoning the food supply of the fungi. The three types of preservative in common use are the *tar-oil, organic solvent* and *water-borne* types.

Tar-oil preservatives are derivatives of coal tar, wood tar or similar extractives, for example coal tar creosote. They are relatively inexpensive and very effective, but have some practical disadvantages such as odour, their liability to stain other materials and the fact that they cannot be overpainted directly with standard finishes.

Organic solvent types consist of various toxic chemicals in an oil solvent which is usually volatile (e.g. white spirit, naphtha or petroleum distillates). They are relatively costly but offer good penetration, dry quickly and can be overpainted.

Water-borne types are aqueous solutions of one or more toxic salts (e.g. copper sulphate, zinc chloride and aluminium sulphate). They are convenient to use, offer fairly good penetration and allow overpainting when dry. However, the re-drying of treated timber by kilning may be necessary. Some are unsuitable for external use owing to leaching tendencies.

Applying preservatives

It is important when applying preservatives to timber to achieve good penetration. Application by brush or spray is less effective than by some other methods. Steeping in preservative (hot or cold, or both in succession) gives better results. More effective are the pressure methods, in which the timber is steeped in the preservative, which is usually hot, inside a sealed tank and different pressures (or vacuum) applied in a certain sequence to force the preservative into the timber.

In another method of preservation, known as the *diffusion impregnation process*, the unseasoned timber is dipped in, or sprayed with, a preservative solution, then close-piled and covered to prevent evaporation. In this way, considerable, or complete, penetration is achieved by diffusion of the preservative throughout the timber. This process cannot be applied after seasoning, but in consequence avoids the necessity of re-drying following preservation.

Natural durability, sapwood and heartwood

Some timbers do not need preservative treatments because they are naturally durable. For example, teak, iroko, jarrah, greenheart, European oak and western red cedar are all renowned for their inherent resistance to decay. Another factor is that heartwood is in general more durable than sapwood, due principally to the presence in the sapwood of sugars and starches which are food for the attacking agency. Also, the penetration of fungal hyphae into the heartwood is often made difficult by the presence of deposits and tyloses which can block entry to the cell cavities. You will realise also that it is easier to impregnate sapwood with preservative than heartwood, so that treated sapwood can be more durable than untreated heartwood.

Insect attack

Preservatives are also effective in preventing the attack of timber by beetles, certain of which cause considerable damage to timber. The typical life cycle of these beetles is a transition through the four stages of egg, larva, pupa and adult (or beetle). The female beetle deposits its eggs at or just beneath the surface of the timber. These hatch into the larvae (plural of larva), which have the appearance of a maggot or grub and are equipped with powerful biting jaws which enable them to tunnel through the timber, feeding on the wood or cell contents and growing to full size in the process. The larva changes to pupa form, an inactive stage during which the change to beetle form is accomplished. The beetle then emerges from the timber leaving an exit hole ('flight' hole) as evidence of its activity.

In general, an attack is favoured by damp conditions and the presence of sapwood or fungal decay. Unpainted timber with rough surfaces is most suitable to receive eggs. The common furniture beetle, the death-watch beetle and the lyctus beetle are all known to fly and can therefore easily carry infestation to new sites. As previously stated, preventive measures include the use of toxic preserva-

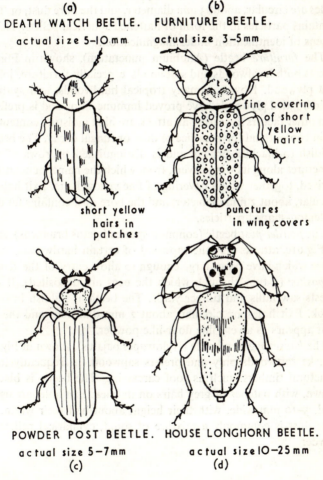

(a)
DEATH WATCH BEETLE.
actual size 5–10 mm

short yellow
hairs in
patches

(b)
FURNITURE BEETLE.
actual size 3–5 mm

fine covering
of short
yellow
hairs

punctures
in wing covers

POWDER POST BEETLE.
actual size 5–7 mm
(c)

HOUSE LONGHORN BEETLE.
actual size 10–25 mm
(d)

Fig 61 Common wood-destroying beetles

tives. Remedial treatments include the use of proprietary insecticides with dipping, spraying or brushing, as well as injection into flight-holes, fumigation and heat treatment. All measures must be thorough, and repeated at intervals, to be effective.

The *death-watch* beetle (Xestobium rufovillosum), shown in Fig 61a, attacks mainly hardwoods, especially old timbers and those subject to decay. The beetle is a chocolate-brown colour, often distinguished by the presence of tufts of short yellowish hairs on its back. Exit holes are circular, about 3 mm diameter, and the bore dust, or 'frass', contains particles in the form of flattened spherical pellets. This is a means of identification under magnification.

The *furniture* beetle (Anobium punctatum), shown in Fig 61b, infests both hardwoods and softwoods, especially furniture, joinery, and plywood. However, many tropical hardwoods, and synthetic-resin bonded plywoods have proved immune. Sapwood is preferred, and other factors favouring an attack are high moisture content, low resin content and the presence of dote or fungal decay. The beetle is reddish to blackish brown, and is distinguished by rows of small punctures along its wing covers, from which the name 'punctatum' is derived, together with a covering of fine yellow hairs. Exit holes are circular, about 2 mm diameter, and the bore dust contains fat cigar- or lemon-shaped particles.

The *powder post* beetle (common species Lyctus brunneus), shown in Fig 61c, attacks only the sapwood of certain hardwoods, usually those which have pores large enough to allow entry of the female's *ovipositor* (a tube through which the eggs are deposited). It often infests saw mills and timber yards. The beetle is reddish brown to black. Exit holes are circular, about 2 mm diameter, and the bore dust appears as a very fine flour-like powder.

The *house longhorn* beetle (Hylotrupes bajalus), shown in Fig 61d, attacks softwoods only and prefers sapwood. It frequently infests structural timbers, such as roof carcassing. The beetle is black or brown, with patches of grey hairs on its back. Exit holes are usually oval, 5–10 mm wide, with their height about half their width. The bore dust is commonly a mixture of squat, cylindrical pellets and powder.

The usual method of grading timber is to assess its general character-istics, as determined by its *growth environment* and by its treatment during conversion, seasoning, storage and handling.

Characteristics of type and species

If the timber is of a known species then the hereditary character-istics or properties will be known, and in conjunction with the assess-ment will enable it to be graded, that is, classified according to its suitability for general and structural uses. In view of the much greater number of species of hardwood than of softwood, and their more complex cellular structure, the grading of softwoods is relatively more simple than the grading of hardwoods.

Grading and moisture content

Since timber will shrink on drying, with changes in its strength and the possible development of defects such as splits, it is important to relate the grading of a sample to its current moisture content.

Grading characteristics and defects

Some of the important characteristics and defects considered in grading will now be listed, together with observations on their significance.

[a] *Slope of grain.* This refers to the inclination of the fibres to the longitudinal axis of the piece of timber. It is important in relation to the strength of a member, and specifications will often limit this. The slope of grain may be determined as shown in Fig 62, using a scribe consisting of a needle fixed in a bent rod with a swivelling handle.

[b] *Rate of growth.* This is expressed as the number of growth rings per inch, or per centimetre, at the ends (cross-section) of a piece of timber (the average for the two ends). This relates to the structural value of the timber, and a minimum number may be specified.

[c] *Wane.* This term is applied to the presence in a piece of timber of the rounded external surface of the original trunk, which limits the maximum finished size of the piece and also indicates sapwood.

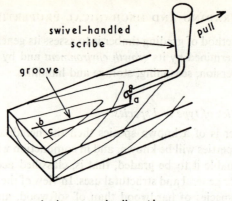

swivel-handled
scribe

groove

pull

a

b

c

ab shows grain direction
ac is parallel to edge of piece
bc is at right angles to ac

slope of grain = one in $\frac{ac}{bc}$

Fig 62 Determining the slope of grain of timber

Its presence in unfinished timber is usually tolerated except in the highest grade of timber.

[d] *Knots and knot holes.* The number, size and disposition of these may be determined by inspecting the piece of wood. The extent to which they are admissible depends on the proposed use of the timber. In structural timbers, the size of individual knots usually needs to be limited in relation to the size of the section of the member. Loose knots are more detrimental than sound knots, but for some classes of work these may be removed and the holes plugged.

[e] *Checks and shakes.* These are cracks or splits caused by the separation of the fibres along the grain, often during seasoning. The assessment is usually by stating the aggregate length of all splits in a piece of timber, and the result can affect the grading.

[f] *Sapwood.* The extent to which this is present is usually stated in terms of the percentage of the width of the piece concerned. Sapwood is usually permissible but may affect the grading. It may influence the use of the timber, since for some purposes sapwood may be unsuitable unless treated with a preservative.

[g] *Stain*. Stain is usually permissible but may affect the grading, and can be important in relation to the proposed use of the timber.

The measurement of characteristics and defects

Information on the assessment of certain of the characteristics and defects listed above are included in BS 1860 (Part 1), 'Structural timber—measurement of characteristics affecting strength'.

Specific requirements and recommendations in relation to these, and other, characteristics and defects of timber are given in various BS publications.

BS 1186 (Part 1) 'Quality of timber and workmanship in joinery' includes a table listing common species of timber, together with notes on their suitability for different purposes.

BS Code of Practice 112, 'The structural use of timber in buildings', gives comprehensive information concerning both the grading of timber and relevant design data.

Tests on clear specimens of timber

Clear specimens are specimens which are free of knots and other defects. BS 373, 'Methods of testing small clear specimens of timber', gives methods of testing timber to determine its mechanical properties. The tests dealt with include: moisture content, density, strength in bending, direct compression, shear and tension, surface hardness (indentation test) and drying shrinkage. Such tests, apart from their routine use for quality control purposes, furnish the initial information necessary to enable design data to be applied to commercially graded timber (see page 212), and can also be the means of assessing the structural characteristics of a *new* species, that is, a species not previously in commercial use.

Standard tests are now made mainly on specimens cut to a 2 cm module; a 2 in module was formerly used. Since the strength of timber varies for moisture content below fibre saturation point (about 25%), the moisture content at which tests are made should be stated. Normally, specimens are conditioned at 20°C and 65% relative humidity before testing. A summary of the principal tests for the structural properties of timber follows.

1 *Compression parallel to the grain* (Fig 63).
The load is applied between parallel plates. Data can be obtained to

allow the calculation of the modulus of elasticity in addition to the compressive stress at both the limit of proportionality and at maximum load.

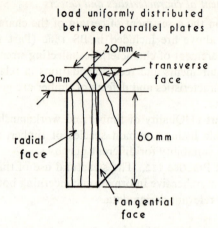

Fig 63 Compression parallel to grain test (timber specimen)

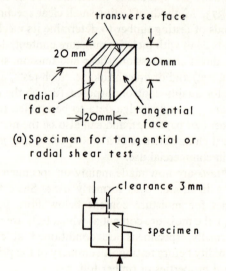

(a) Specimen for tangential or radial shear test

(b) Method of loading

Fig 64 Shear parallel to grain test on timber

2 Shear parallel to the grain

Tests are made as indicated in Fig 64, both for shear along the tangential plane and along the radial plane.

3 Cleavage test

Tests are made as shown in Fig 65, both for radial and tangential

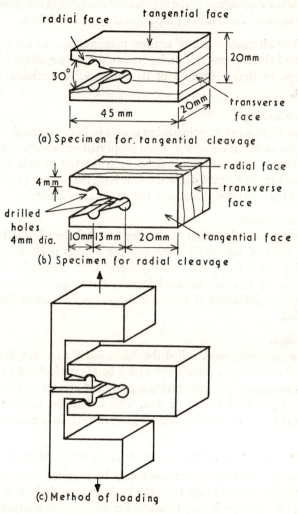

(a) Specimen for tangential cleavage

(b) Specimen for radial cleavage

(c) Method of loading

Fig 65 Cleavage test on timber

cleavage, and the result is stated as the failing load per centimetre of width (i.e. cleavage).

4 *Tension parallel to grain*
Fig 66 shows the form of test piece used, which is axially loaded through toothed grips applied to its ends. Data can be obtained which allow calculation of the modulus of elasticity in addition to the tensile stress at the limit of proportionality and at failure.

Note The alternative test of tension perpendicular to the grains is little used since in practice designers avoid applying direct stress of this type to timber because of its comparative weakness in this respect.

5 *Static bending*
The test piece is centrally loaded as shown in Fig 67 through a bearer of 30 mm radius of curvature, when supported on bearers at 280 mm span. Data which can be obtained from the test include the fibre stress and the horizontal shear stress at both the limit of proportionality and at maximum load, also the modulus of elasticity.

6 *Impact bending*
The arrangement is similar to that for the static bending test except that a span of 240 mm is used and the load is applied by a tup weighing 3·3 lbf (1·5 kgf) falling freely from successive heights, increasing by specified regular intervals. The height of fall to cause failure, or a deflection of 60 mm if this is reached before failure, is recorded.

7 *Hardness*
The test normally used, called the *Janka indentation test*, is to find the load necessary to force a bar with a hemispherical end or a steel ball of diameter 11·28 mm giving a projected area of 100 mm² to a depth of penetration of 5·64 mm. This is half the diameter. The hardness of both the radial and tangential surfaces is measured.

Relating test data to commercial grading, for design purposes
The designer of timber structures may be required to use *design* (or *working*) stresses lower than the basic stresses derived from the testing of clear specimens, to allow for any adverse characteristics and defects shown to be present by the grading.

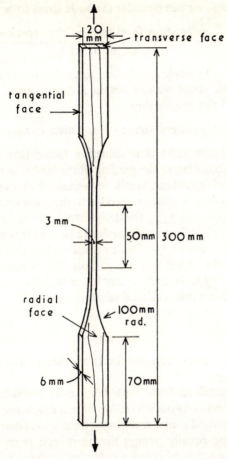

Fig 66 Tension parallel to grain test on timber

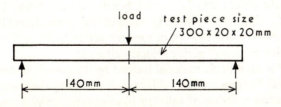

Fig 67 Static bending test on timber

For simplicity, we can consider the basic stress to be given by:

$$\text{Basic stress} = \frac{\text{mean stress from tests on clear specimens}}{\text{factor of safety}}$$

(In practice more complex, statistical methods may be used.)
From the basic stress we can obtain a design stress. One simple procedure uses the relationship:

$$\text{Design stress} = \text{basic stress} \times \text{strength ratio}$$

where the strength ratio is a reduction factor (less than 100%) which varies according to the grading of the timber and the type of stress concerned (e.g. shear, tensile or compressive). For example, if the basic stress for compression parallel to the grain were 80 kgf/cm² and the strength ratio 75%, the design stress would be 60 kgf/cm².

The need to allow for varying moisture content is usually avoided by using basic stresses applicable to *green* timber, that is, timber at the fibre saturation point. For this purpose the conversion of a basic stress obtained by tests made at any lower moisture content can be made using known mathematical relationships.

Summary

Timber is useful structurally and for its appearance and comparative ease of working.

Timber is obtained from both hardwoods (broadleaf trees or angiosperms) and softwoods (conifers or gymnosperms).

Both the sapwood (outer trunk) and heartwood (inner trunk) are useful. Both are equally strong; the heartwood is more naturally durable but the sapwood is more easily impregnated with preservative.

A growth ring is one season's growth of wood, forming initially a layer at the outer part of the trunk beneath the cambium. The inner part of the ring constitutes early wood or spring wood, and the outer part constitutes the late wood or summer wood which is usually darker due to slower growth giving thicker cell walls.

Wood is an organic material formed by the growth of cells in the living tree. The tree produces this organic matter by the conversion of sap (aqueous solution of salts) to foodstuffs (carbohydrates) in the leaves. The process is known as photosynthesis.

Foodstuffs are consumed in the cambium to form living cells containing protoplasm, which subsequently modify and lignify to form the different classes of wood cells.

Wood cells are composed mainly of cellulose and the cementing agent lignin.

Softwoods have only two types of cell, the pod-like tracheids (forming mechanical and conducting tissue) and brick-shaped parenchyma cells (providing food storage).

Hardwoods have needle-like fibres (mechanical tissue) and tube-like vessels (conducting cells), in addition to parenchyma and sometimes tracheids.

Pits are unthickened parts of cell walls, permeable to liquid solutions in the living tree, and which may connect like or unlike cells. Bordered pits differ from simple pits by the valve-like action of their torus pad. They occur only in tracheids.

The botanical classification of timbers is by family (e.g. pine or *Pinacea*), genus (e.g. pine or *Pinus* and spruce or *Picea*) and species (e.g. Scots pine or *Pinus sylvestris*, Canadian yellow pine or *Pinus strobus*, and American long-leaf pitch pine or *Pinus palustris*).

Hardwoods are usually identified by their general appearance, colour, texture, density and smell, together with a hand lens examination of features such as rays (width), pores (ring or diffuse porous, size, shape and arrangement) and presence or absence of tyloses or deposits.

Softwoods are often identified by their general appearance, colour, texture, density and smell, together with features such as the number of growth rings and the thickness and intensity of the late-wood band, the presence or absence of resin ducts (holes in a cross-section, or dark streaks along a tangential face). For positive identification, the three principal sections must be examined under a microscope.

Conversion is the preliminary cutting of the felled tree and its subsequent reduction to baulks and planks before seasoning.

Seasoning is the controlled reduction of moisture content of the converted timber to a level below the fibre saturation point and approximating to the required final equilibrium moisture content.

Timber may be air-seasoned to relatively high moisture contents, or kiln seasoned more quickly to high or low moisture content.

Timber dried below fibre saturation point undergoes shrinkage which is reversible by wetting or hygroscopic action, and acquires increased strength but reduced resilience.

Durability

Fungi attack timber as a source of food. Their other requirements are oxygen (air), moisture and favourable temperature conditions. All are sterilised by kilning but this does not prevent future attacks.

Attack is spread by microscopic spores which float in the air, or by the spread of hyphae from an existing growth.

The many types of fungi include the wood rots, which destroy the cell walls, and surface moulds and sap staining fungi which feed only on the cell contents.

The wood rots include brown rots, which attack only the cellulose of the cell walls (e.g. the dry rot fungus *merulius lacrymans* and the cellar rot *coniophora cerebella*), and the white rots, which attack cellulose and lignin (mainly in the forest and fencing).

Impregnated preservatives act by poisoning the fungi's and wood beetles' food supply because they are toxic, that is, poisonous.

Preservative types are tar-oil, organic solvent and water-borne.

Application is by brush, spray, steeping (hot or cold), pressure or vacuum methods, or diffusion impregnation.

The life cycle of the common wood-destroying beetles has four stages (egg, larva, pupa and adult).

The larva ('grub') causes damage by tunnelling through the wood, and the emerging beetle leaves a flight hole.

The chocolate-brown death-watch beetle attacks mainly old hardwoods, leaving a circular flight hole, and bore dust containing bun-shaped pellets.

The similar-looking, but smaller, reddish-brown furniture beetle attacks both hardwoods and softwoods, leaving a small circular flight hole and bore dust containing cigar- or lemon-shaped particles. It can be identified by its punctured wing covers.

The common powder-post beetle attacks only the sapwood of certain hardwoods, notably in timber yards, and leaves a small flight hole and flour-like bore dust.

The house longhorn beetle attacks softwoods only (e.g. in roof carcassing) and prefers sapwood. It leaves large oval flight holes and bore dust of cylindrical pellets and powder.

Grading and mechanical properties

The testing of clear specimens of timber provides data from which basic stresses (incorporating a factor of safety) are derived.

Commercial grading classifies timber according to type (hardwood or softwood) and species, physical characteristics and defects to assist the user, designer and specification writer.

The designer modifies the basic stresses according to the grading of the timber (e.g. by the 'strength ratio' method).

Characteristics assessed in grading at a stated moisture content include slope of grain, rate of growth, wane, knots and knot holes, checks and shakes, sapwood and stain.

Tests on clear specimens include compression, tension and shear parallel to the grain, static and impact bending, hardness and cleavage.

12

PLYWOOD, BOARD AND SLAB MATERIALS

PLYWOOD

In spite of the fact that there is a tendency for the term plywood to be applied collectively to several forms of timber board (see Fig 68), for classification and to avoid ambiguity in specification a distinction must be made between plywood of the traditional type and products such as blockboard and laminboard,

Plywood is a timber product consisting of a number of thin sheets or *veneers* of wood bonded together with an adhesive. Finished thicknesses vary from about 3 to 25 mm. The important characteristic of plywood is the vastly improved strength characteristics which result from bonding alternate veneers with the grain aligned in different directions, normally at right angles. Another benefit obtained by alternating the grain direction is a reduction in the maximum movement due to moisture changes, since the greater movement across the grain is restrained by the very small movement along the grain of adjacent veneers. A balanced construction, using an 'odd' number of veneers (3, 5, 7 and so on), is usual, although not invariable, and reduces any tendency for distortion. Boards with more than three veneers are termed *multiply*.

The use of specially selected or expensive timbers for the surface plies only is economical. Veneers are always sorted for quality before they are used so that different grades of plywood can be manufactured to suit different purposes. The natural durability of the plywood must necessarily depend not only on the species of the veneers, but also on the type of adhesive used.

During manufacture, selected *boles* (trunks) are normally first softened by soaking, boiling or steaming, and then peeled on a rotary lathe to obtain a wide veneer of any length (1 to 4 mm thick). In

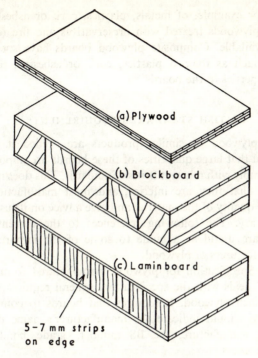

(a) Plywood

(b) Blockboard

(c) Laminboard

5-7 mm strips
on edge

Fig 68 Plywood, blockboard and laminboard

this way, boards of any size can be manufactured, which is a very important consideration to the user. The procedures of the remaining stages of manufacture—gluing, assembling, pressing and final conditioning—are fairly standard, but the amount of initial conditioning of the veneers before gluing varies. In general, the best quality boards are obtained by carefully drying the veneers to between 5 and 15% moisture content before gluing. Gluing the 'wet' or 'semidry' veneers produces boards of lower quality which can cause imbalance in the finished board, with possible warping, checking or splitting. In all cases, the boards should be finally conditioned to about 12% moisture content before dispatch.

Faced plywood

Plywoods are obtainable with bonded facings of insulating, sound absorbent, fire-resistant, abrasion-resistant or purely decorative

220   SCIENCE IN BUILDING

sheeting (for example, of metals, plastics, cork or asbestos). Impregnated plywoods treated with preservatives and fire retardants are also available. Composite plywood boards have special core materials, such as foamed plastics, cork or asbestos, to impart specific properties to the boards.

BRITISH STANDARDS REQUIREMENTS

Whenever plywood or similar products are needed it must be remembered that large quantities of these boards are imported and may not comply with relevant British Standards. This does not, however, mean that they are inferior, and most manufacturers will support the quality of the material and give advice on its uses whenever necessary. The following references to the current British Standards are useful as a guide to some of the important considerations for users of plywood.

BS 1455 'Specification for plywood manufactured from tropical hardwoods' deals with the specification, general requirements and testing of such plywood. It requires that boards to comply with the standard should indicate the manufacturer's name or mark, country of manufacture, the BS number, grade bonding and nominal thickness.

Grade refers to the quality of the surface veneers of the plywood, and bonding to the class of adhesive used, according to its durability (see Tables 34 and 35 pages 323–4). The grade of the surface veneers for any board is designated by the use of two numbers (for example, 1–3), each of which refers to one surface veneer. Brief notes of the tests given in BS 1455 follow.

1 *Moisture content* This is determined by oven-drying a small specimen

2 *Bond test* The *knife test* is made on wet test pieces previously subjected to steaming, boiling or cold immersion as determined by the bonding designation. A special form of knife, designed for use with a chisel action, is pushed into the veneer with the blade parallel to the grain, and a forward thrust is applied along the glue line. The degree of separation of the veneers achieved is assessed according to a master 'points' scale of 1 to 10 (see BS 1455). An average result is obtained for a number of specimens

3 *Mycological test* (for plywood of bonding designations WBP, BR

and MR only—see Table 35). This test is to assess the resistance of the plywood to attack by micro-organisms (see Experiment 31)

EXPERIMENT 30 The effect of immersion on the bonding of plywood

Note The knife test for bond to BS 1455 could be included in this experiment

Apparatus Large dish

Specimens Plywood test pieces size 200 × 100 mm (internal and external grades)

Method
1 Immerse the specimens in water at room temperature for between 16 and 24 hours, allowing access of water to all surfaces
2 Remove the specimens from water and examine them immediately for signs of separation of the plies at the edges, or the sliding of one ply over another

Results Record your observations

Supplementary work Another set of specimens may be tested for their resistance to boiling water. Boil for three hours, rinse in cold water and check the bond again.

EXPERIMENT 31 Mycological test on plywood

Apparatus Dish (glass, porcelain or enamelled iron) for specimens, with sheet glass cover—measuring cylinder (500 ml)—large beaker—low-temperature oven (24 ± 2°C)—thermometer 0–50°C—sawdust obtained from the sapwood of any timber, or from perishable timber in its natural condition (for example, ash, beech, birch, poplar or willow)—sugar solution (13 g of table sugar to 1 l of water)—modelling clay or linseed oil window putty

Specimens Two plywood test pieces (100 × 50 mm)

Preparation
Moisten the sawdust with sugar solution. Use about three times its weight of solution to the dry sawdust, and place it loosely compacted in the dish to form a layer 25 mm deep. Cover and seal the dish to prevent evaporation and keep it in the oven at 25°C for one week.

Over the same period immerse the test pieces in cold water,

changing the water daily to remove any free formaldehyde or other mould-inhibiting substance

Test procedure

1 Remove the cover from the dish and press the wet test pieces into the sawdust so that their upper surface is level with that of the sawdust.

2 Replace and reseal the cover, then leave the dish at $25 \pm 2°C$ for five weeks.

3 Remove and wash the specimens in cold water, then examine them for separation of the plies, blistering or other failure of the glue line

Results Record your observations

BLOCKBOARD

Blockboard is a board consisting of a core of wood strips about 25 mm wide sandwiched between outer veneers laid and glued with the grain at right angles to the core strips (Fig 68b). The strips, usually of softwood, may be left unglued (*stripboard*) or may be glued together. The construction described is of three-ply (core and single veneers), but five-ply (core and double veneers) is also made. The finished thickness is from about 15 to 30 mm. This product is dealt with in BS 3444.

LAMINBOARD

Laminboard is of similar construction to blockboard but it has core strips, 3–7 mm wide, glued together (Fig 68c). The boards are usually heavier than blockboard, due to the greater amount of adhesive used, and, in some cases due to denser timbers, often hardwoods. The product is dealt with in BS 3444.

PARTICLE BOARDS

Particle board, formerly called wood chipboard, consists of size-graded wood chips bonded together, usually under pressure and heat, by an adhesive which is normally a thermosetting synthetic resin. Adhesives of the urea formaldehyde type are often used, in which case the board is not suited to external use. Boards of different thicknesses up to about 40 mm are obtainable in large sizes.

Lower density boards are sometimes produced by an extrusion process instead of mechanical pressing, and these may be used to form a *coreboard* by applying outer veneers of timber or other sheet materials.

Boards with various facings and applied finishes are obtainable. BS 2604 deals with resin-bonded wood chipboard.

WOOD WOOL SLABS

Wood wool slabs are made by shredding timber into long fibres and binding these together with a cementing agent, usually ordinary Portland cement. The open-textured surface offers good sound absorption if left unsurfaced, or alternatively, provides good key for plaster and rendering.

Slabs are made to various sizes, and are between 30 × 100 mm in thickness. Special heavy duty grades and reinforced slabs are available to meet structural requirements.

STRAWBOARD

Strawboard boards consist of compressed straw between millboard (a paper product) or other sheet materials and are usually about 50 mm thick. They can be supplied specially surfaced to receive plastering. The material normally has a high moisture movement, and should not be used in damp conditions.

FIBRE BUILDING BOARDS

Fibre building boards are made by pulping fibrous materials of vegetable origin, mainly timber, straining to remove excess water, then pressing the mass with applied heat to cause binding of the fibres. The binding action may result from the lignin content of the wood alone, or other bonding agents may be added.

Boards are commonly classified as insulating boards, wallboards or hardboards, mainly according to their density and thickness which govern their normal uses.

Insulating boards

Insulating boards are low density boards, up to 25 lb/ft³ (400 kg/m³), and not less than 10 mm thick. The three main types are [a] *homogeneous insulating boards,* used for their good thermal insulation, sound absorbency and anti-condensation properties, [b] *bitumen bonded* types, highly resistant to moisture, and [c] *bitumen impregnated* types, which are either grey (lightly impregnated) or black (heavily impregnated, for example, for expansion jointing).

Wallboards

Wallboards are very similar to insulating boards but may be slightly heavier, up to 30 lb/ft³ (480 kg/m³) and are thinner, offering greater economy. *Homogeneous types* ('building board') are usually 8–10 mm thick, and *laminated* types are about 5 mm thick.

Hardboards

Hardboards come in three grades, according to density. *Standard* hardboards, over 55 lb/ft³ or 880 kg/m³, 3–5 mm approximately in thickness are a general purpose building board which is widely used. *Medium* hardboards, 30–50 lb/ft³ (480–800 kg/m³), 5–12 mm thick, give better thermal insulation than standard boards but have lower strength and abrasion resistance. *Tempered* hardboards (density over 60 lb/ft³ or 960 kg/m³) are tough, wear-resistant and highly weather-resistant.

Special types of these hardboards are available with different facing materials or surface treatments. Some are impregnated for fire resistance or perforated for sound absorbency and for use as pegboard.

Composite types of board may contain a core of foamed plastics, cork or glass fibre.

BS 1142 deals with fibre building boards.

CORKBOARD

Corkboard is made by compressing granulated cork and applying steam heat. This 'baking' causes the natural resin of the cork, obtained from the outer bark of the cork oak (*Quercus suber*) and known as corkwood, to cement the particles firmly together. Cork is a low density, micro-cellular material and consequently has excellent thermal insulating properties. It is unique in that it shows no capillarity, and is consequently very suited for use in refrigeration work because it does not absorb condensation. It can also be used as a flooring material as its texture deadens sound, and it also provides a decorative finish, either plain or patterned.

ASBESTOS-CEMENT SHEETS

Asbestos is a natural fibrous mineral, a silicate of magnesium or iron, which is light in weight yet has good longitudinal strength. It is a good insulator, is incorrodible and highly fire resistant. These qualities make it suitable for use as a reinforcing material in conjunction with Portland cement. The moist mixture is initially pressed into flat sheets but these may subsequently be moulded before hardening occurs to form curved or corrugated sheets if required. Standard mixes contain 15–20% of asbestos, the remainder being cement, and produce three grades of *flat sheet*, [a] *semi-compressed*, for ceiling and wall linings where condensation is not a risk; [b] *fully compressed* with greater strength and smoother surfaces; for wearing surfaces (for example, bench tops), bath panelling and cladding, [c] *flexible compressed* (3–4 mm thick), which can be curved to small radii.

Other sheets are made containing more asbestos fibre. BS 690 deals with corrugated and semi-compressed flat sheets. Fully compressed flat sheets are dealt with in BS 4036.

Asbestos wallboard

Asbestos wallboard is a sheet highly suitable for wall and ceiling linings.

Asbestos insulating board

Asbestos insulating board is a low density board for thermal insulating purposes. It is also used to resist high humidities, as it is less prone to condensation than normal density types, and high temperatures (unlike types [a][b] and [c] above, which can shatter at extremes of temperature).

H

Asbestos partition board

Asbestos partition board differs from other sheets in that it contains some organic fibre in addition to asbestos, and is classified as combustible.

Danger with asbestos cement

The reinforcement value of asbestos alone is quite inadequate to allow the application of heavy loads to asbestos-cement sheets, and the brittleness of this material makes it highly dangerous for persons to walk, stand or lie on the sheets.

Summary

Plywood consists of a number of veneers of wood bonded together with an adhesive, normally with the grain direction of alternate plies at right angles. This increases strength and reduces movement.

Manufacturing stages include softening treatment, peeling of veneers, their initial conditioning to dry, wet or semi-dry, gluing, assembling, pressing and final conditioning.

Large sheets can be made, if necessary, using facing veneers of selected grade or special facing materials (metals, plastics, cork or asbestos). Sheets impregnated with preservative or fire-retardant solution are obtainable.

Tests for plywood are for moisture content, bond (knife test) and resistance to micro-organisms.

'Bonding' refers to the class of adhesive used, with respect to its moisture resistance, which is tested by immersion in cold or hot water, or steam.

Blockboard consists of core strips about 25 mm wide glued between outer veneers (15–30 mm thick).

Laminboard is heavier than blockboard with strips 3–7 mm wide glued together and between outer veneers.

Particle boards are of compressed and bonded size-graded wood chips, in board thicknesses up to about 40 mm.

Coreboards have a core of particle board, which may be extruded for low density, bonded to outer veneers.

Wood wool slabs consist of shredded timber bonded with ordinary Portland cement (30–100 mm thickness).

Strawboards are of compressed straw between millboard or other sheets.

Fibre building boards are made mainly by pulping and straining wood fibre, then compressing it, with applied heat. Types are [a] insulating boards (low density, not less than 10 mm thick)—homogeneous, bitumen-bonded or impregnated [b] wallboards (medium density less than 10 mm thick) homogeneous or laminated [c] hardboards (medium, standard or tempered). In addition, there are special types (faced, impregnated or perforated) and also composite boards.

Corkboard is compressed granular cork in sheet form and is useful for its low density, high thermal insulation and resilience, together with the absence of capillarity.

Asbestos-cement flat sheets are of three types, *semi*-compressed, *fully*-compressed and *flexible*-compressed. Other types are asbestos wallboard, and the heat and moisture resistant (anti-condensation) asbestos insulating board.

Asbestos-cement sheets are brittle and easily fracture unless specially reinforced with steel.

13

PLASTICS

Plastics are a group of materials which harden during manufacture from a condition in which they can be moulded to any desired shape. Plastics materials are typically of *organic* composition and are formed by a process of molecular growth known as *polymerisation*.

Organic substances

These are compounds principally of carbon and hydrogen, often together with other elements, notably oxygen, nitrogen, sulphur and the halogens (fluorine, chlorine, bromine and iodine).

Polymerisation

Polymerisation is the linking together of the like molecular units, termed *monomers* to form a chain, or grid-like, structure, the *polymer*. The polymerised substance is liable to possess vastly different physical and chemical properties to those of the monomer. In fact, it is possible to vary the final properties by controlling the size and arrangement of the molecular chains during manufacture. For example, increasing the length of the 'chains' results in a higher softening temperature and usually greater strength.

The process of polymerisation is usually brought about by subjecting the raw material (the monomer form) to the action of one or more such influences as light, heat, pressure or a *catalyst*, a substance which promotes a chemical reaction. Whereas these plastics materials are synthetic, this is not necessarily so with the monomers, as some exist (in polymerised form) in naturally occurring substances. for example, in cellulose and starch, in protein compounds, in rubber latex and in natural resins (e.g. shellac, copal and rosin).

Cellulose is the substance which forms the main structure of the cell walls of plants and is a polymer of *glucose*. For commercial use, it is obtained from the cotton plant, which is the purest source, from wood and from straw. It is classified chemically as a *carbohydrate*, which means that it is of molecular composition carbon, hydrogen and oxygen only.

Starch, like cellulose, is based on the *glucose* monomer, and is present in plants as a source of nutriment.

Proteins are a group of organic nitrogenous compounds in polymerised form, which make up the main structure of all plant and animal tissue. They are based on monomers of certain *amino acids*. Two sources of protein, important in the early days of the plastics industry, are casein, which comes from milk, and the soya bean.

Rubber latex is a tree extract based on a monomer called *isoprene*.

Copal, *rosin* and *rubber latex* are extractives of certain species of trees, and *shellac* is an extractive of certain insects.

All these naturally occurring polymers—cellulose, starch, proteins, rubber latex and resins—have what is described as *straight-chain*, or linear, molecular construction, which means that the monomers are linked end to end in a single row. The polymer is simply made up of a series of these chains lying together with their length parallel to a common axis, but with their ends overlapping. With this picture in mind, it is obvious that the ease with which these 'chains' (the polymers) can be caused to slide over one another will depend to a great extent on the length of the individual chains and the amount of *twist* (the angle between alternate 'links' of each chain) in them. Such features profoundly influence the strength, flexibility and resistance to heat and moisture absorption of the plastics material.

Thermoplastic and thermosetting plastics

The straight-chain polymer plastics are softened by heat and will reharden on cooling, that is, they are *thermoplastic*, and may therefore be reshaped or remoulded. This can be a useful property in the manufacture and working of the plastics materials.

Other plastics are *thermosetting*, that is, once formed they cannot be re-softened by heat. The typical molecular arrangement of the thermosetting plastics is with the monomers arranged to form a two- or three-dimensional, *grid-like* structure, or with chains *cross-linked*. In general, the *thermosets* are more heat-stable and more resistant to solvent action than the *thermoplastics*.

A co-polymer is a substance obtained by the polymerisation of a mixture of monomers which are then said to be *co-polymerised* and this provides an important means of adding to the range of plastics materials obtainable with different properties.

Other ingredients

Apart from the monomer base, other materials which may be incorporated during the manufacture of a plastics material to serve some specific purpose include solvents, plasticisers, fillers and colouring agents.

Solvents are used to give the material greater plasticity during the manufacturing stage, usually to facilitate moulding to the desired shape. Examples are acetone and chloroform.

Plasticisers are usually non-volatile organic solvents of high boiling point, added to reduce brittleness or increase flexibility of the finished plastics material.

Fillers are added to modify one or more properties of the finished plastics material; for example, as a reinforcing agent to improve strength (e.g. glass fibre) or toughness (e.g. woodflour) or to confer heat resistance (e.g. asbestos). *Colouring agents* may be either dyes or pigments.

Present-day sources and properties of plastics

The majority of plastics in everyday use are made from derivatives of coal-tar, mineral oil (petroleum) and natural gases. The original plastics materials were based on monomers provided by Nature.

Plastics with a wide range of different properties are available. In general, plastics are resistant to water and chemicals, but are softened and expanded by organic solvents. Many thermoplastics have comparatively low softening points and high thermal expansion coefficients. They are easily moulded to complex shapes in manufacture, e.g. by casting, pressing, extrusion or injection moulding, and most can be foamed or expanded to lightweight cellular form. They provide products which are clean, comparatively light and easy to fix and self-finished in many cases.

Cellulose plastics

Celluloid was the first plastics material ever made. It was patented as Parkesite in 1864, and is one of a number of plastics based on cellulose. It is a thermoplastic, made by interacting purified cellulose with a mixture of nitric and sulphuric acids to form cellulose nitrate (nitro-cellulose), with the addition of a plasticiser (e.g. camphor). The process demands carefully controlled conditions because the product is highly inflammable, and in its gun-cotton form is dangerously explosive. It is not, however, dangerously explosive in its celluloid form. The material was originally described as artificial horn, and was particularly useful in its transparent form. It has for most purposes been largely superseded by non-flammable *cellulose acetate* which has similar properties, formed by interacting cellulose with acetic acid and acetic anhydride together with a catalyst. This material finds a multiplicity of uses, mainly in small articles such as door furniture and fittings generally. It is useful for its clarity, when unpigmented, and its toughness, but it expands when exposed to moisture.

Later developments of similar products with greater moisture resistance include *cellulose proprionate* and *cellulose acetate butyrate*.

Casein plastics

These early plastics are made by immersing the thermoplastic casein in *formalin* (a solution of the gas formaldehyde in water), which cures (hardens) it by causing cross-linkage between the polymers. The curing process, known as *formalising*, is rather lengthy. These plastics are little used in the building industry.

Phenolic resins

The name *bakelite* was given, in 1907, to the first fully synthetic plastics material to be made, by the interaction of *phenol* (carbolic acid) with *formaldehyde*. Both substances are obtainable from coal-tar. *Phenol formaldehyde*, often called PF resin, or simply 'Phenolic', is thermosetting.

Phenol is the junior member of the *phenol group* of compounds, which also includes *cresol* and *resorcinol*. Any plastics made by the interaction of a phenol with formaldehyde (or other aldehyde) is classed as a *phenolic resin*. Phenolic resins are widely used for fittings and adhesives. They have high heat resistance but are not highly resistant to impact, unless specially reinforced.

Amino plastics

Formaldehyde monomer is also interacted with urea to form *urea formaldehyde* (UF) and with melamine to form *melamine formaldehyde* (MF), both thermosetting plastics. The fact that both contain the *amine* radical (NH_2) in their molecular formula results in their being called *amino plastics*. They are obtainable in many attractive colours, and apart from other uses are popular as *laminated plastics* sheet, which consists of multi-layered plastics-impregnated paper or fabric bonded under heat and pressure. Melamine resins are renowned for their resistance to abrasion and heat, and are therefore used for wear-resisting surfaces. Both are used as adhesives.

Certain other resins of the amino plastics group are made by using *thiourea* instead of urea. Thiourea differs by having an atom of sulphur in its molecule.

Vinyl plastics

These are thermoplastics based on acetylene gas, and polyvinyl chloride and polyvinyl acetate are the typical examples.

Polvinyl chloride, or PVC, made from the vinyl chloride monomer, is produced in both rigid (unplasticised) and flexible (plasticised) forms, and is widely used in the building industry. The *plasticised* form is a tough, rubber-like material with good chemical resistance and non-flammable, useful as a waterproof membrane and as floor and wall covering in its tiles and roll form. The unplasticised form is used for roofing sheets, rainwater goods, plumbing fitments and pipes although there is some limitation due to its low softening point (70–80°C).

Polyvinyl acetate, or PVA, made from the vinyl acetate monomer, can be emulsified in water for use as a paint base or adhesive, and is also mixed in with concrete to give it resilient properties. It can also be added to plastering, rendering and floor toppings to improve their bonding properties.

Polystyrene

Styrene monomer is made by reacting benzene with ethylene. The polymerised form, known as polystyrene, has also been called polyvinyl benzene, because it is structurally similar to the vinyl resins. This thermoplastic is a naturally transparent solid with good resistance to water and chemicals, and although it is somewhat brittle, it can be modified in manufacture to tougher forms. It is perhaps best known in the building industry in its expanded (lightweight) form, when it is used as insulating sheets and for wall and ceiling tiles.

Polythene

Polythene, or *polyethylene*, is made by heating ethylene gas under pressure, which causes polymerisation of the ethylene monomer. It is a thermoplastic, popular for pipes and plumbing fittings and as a damp-proof and curing membrane and weather-shield.

Polythene for general purposes is available in the density range 0·92–0·96, and its rigidity and softening point increase with density. Low density polythene is softened by water near the boiling point and so cannot be used for hot water services. In its natural, translucent, form it degrades in sunlight as embrittlement and cracking are caused by ultra-violet rays, and it is consequently often heavily pigmented with carbon black in manufacture to prevent this happening. It has a relatively high coefficient of thermal expansion which varies with density but has advantages in its flexibility which prevents frost damage, lightness (it floats on water) and corrosion resistance.

Polypropylene

Polypropylene may be regarded as a close chemical relative of polythene. Based on the propylene monomer, its important differences are a higher softening temperature (135°C) and greater resilience.

Polymethyl methacrylate (e.g. Perspex)

Polymethyl methacrylate is classified as an *acrylic resin*, since the monomer (methyl methacrylate) is an acrylic acid derivative. It is important because of its high degree of transparency to light. It is used for window glazing and as corrugated roof-light sheets, tiles and panelling, as well as for lighting fittings, some plumbing fittings when

reinforced, and as a base for a solvent-type adhesive. It is a thermoplastic.

Nylon

Nylon is a thermoplastic copolymer, which has phenol and ammonia
as raw materials, and is chemically a super (long-chain) *polyamide*.
The length of the chain can be varied to produce grades of nylon with
different properties, particularly with respect to heat resistance and
strength; which are greater than for most other thermoplastics.
Softening points can exceed 200°C. Nylons are self-lubricating and
tough, and are therefore useful for pulleys and door runners.

Polyester and alkyd resins

These plastics are usually thermosetting, but some forms are thermoplastic, depending on the form of the monomer. They are made by
the interaction of certain organic acids with alcohols. Polyester resins
are used in laminated construction, with glass fibre where high
strength is required; for example, as reinforced translucent sheeting.
One class of polyesters, known as *alkyd resins*, is popular as a paint
medium.

Epoxy resins

Epoxy resins, made from certain petroleum derivatives, are extremely
tough, with high heat and chemical resistance. They are used mainly
as adhesives, in paints and as a binder in rendering and abrasion-
resistant floor surfacing.

Polyurethanes

Polyurethanes are a group of plastics of complex chemical forms.
They have a number of specialised uses and can provide a hard,
durable finish in paints, or a flexible mastic, and can also be foamed
to give either a resilient or a rigid lightweight material.

Synthetic rubbers

Synthetic rubbers are usually included in the class of flexible products
known as *elastomers* because of their special elastic properties.
These elastomers are of particular importance as ingredients of
mastics, adhesives and paints.

It is possible to synthesise *latex* (the natural raw material of rubber obtained by tapping certain species of tree) by the polymerisation of *isoprene*, the monomer of which it is composed. In addition to *polyisoprene*, there are a number of other synthetic 'rubbers', each having slightly different characteristics (e.g. of softening point, hardness, flexibility, durability, flammability, chemical resistance and slipperiness when wet). *Neoprene* (polychloroprene) is a near relative of polyisoprene. *Buna* rubbers are polymerised butadiene monomer.

There are also various *copolymers*, such as *nitrile rubber* (acrylonitrile-butadiene), *butyl rubber* (*iso*-butylene-butadiene), *styrene-butadiene* and *ethylene-propylene*. Apart from their more conventional uses in sheet form, certain rubbers are used as latex in its emulsified form in water mixed with Portland or other cement and sand to provide resilient *latex-cement* flooring.

Silicones

These form a distinctive group of polymeric substances based on both silica (silicon dioxide, SiO_2) and carbon compounds. The range of products from this source is considerable, and continues to increase. Important products are the silicone waterproofing and fire retardant solutions, silicone resin paints and mould-release agents, silicone elastomer sealants and silicone rubbers. These products all resist extremes of temperature and are water repellent.

The preceding list of plastics represents just a few of the many synthetic plastics materials in use at the present time.

TESTS ON PLASTICS

A considerable number of plastics materials are dealt with by various British Standards and it is proposed to mention only two of these here, to indicate some typical properties which are assessed.

BS 3260 'PVC (vinyl) asbestos floor tiles' deals with tiles composed of a uniform blend of thermoplastic binder (vinyl chloride polymer and/or vinyl chloride copolymer), asbestos fibre, fillers and pigments.

BS 3261 'Flexible PVC flooring' deals with both roll and tile forms of either uniform or laminated composition based on thermoplastic binder (vinyl chloride polymer and/or vinyl chloride copolymer), fillers and pigments.

Tests are given and limits specified in BS 3260 for *colour fastness* (the tendency to fade in daylight); *dimensional stability*, expressed as the percentage change in dimensions when heated from 23 to 80°C and recooled; *content of volatile material*, as loss in weight on heating to 100°C (maintained for 6 hours), with initial and final weights determined at room temperature; *curling* (of 7 in/150 mm square specimens laid flat standing in water, for 72 hours); *indentation* (the measured penetration of a 0·25 in/6·35 mm diameter rod with an hemispherical end for an applied load of 30 lbf/13·61 kgf), relative to an initial load of 2 lbf (0·907 kgf), at stated short-term intervals and temperatures; *residual indentation*, a test similar to that for indentation but using a flat-ended rod, a heavier load and a longer period of loading; *deflection* (of 9 × 2 in/225 × 50 mm test pieces with 3-point loading through ¼ in/6·35 mm diameter steel rods, with supports 8 in/205 mm apart, tested both along and across the grain); *impact resistance* (to a 1 in/25·4 mm diameter steel ball dropped from a height which depends on the thickness of the specimen); *resistance to various substances* (test pieces, 3 × 2 in/75 × 50 mm) are subjected to a standard scratch test and examination for colour change after immersion for 46 hours at room temperature in 95% ethanol, light mineral oil, vegetable oil, 2% sodium hydroxide aqueous solution and beef tallow, the latter being initially warmed to allow immersion.

Comparable tests are specified in BS 3261.

Summary

Plastics materials are a group of polymeric substances, typically organic and synthetic.

During manufacture, or application, the monomer units are polymerised or copolymerised by the application of heat or pressure, or by the use of a catalyst. Other ingredients may be solvents, plasticisers, fillers and colouring agents.

Thermosetting plastics (not re-softened by heat) have the monomers arranged in a grid-like structure.

Straight-chain polymers are typically thermoplastic (re-softened by heat), but if cross-linked are thermosetting.

Naturally occurring polymers include cellulose, starch, protein (e.g. casein and soya bean), rubber latex and resins.

Cellulose (a polymer of glucose) is used to make the cellulose plastics, e.g. the thermoplastics cellulose nitrate (celluloid) and cellulose acetate.

Casein plastics are made by formalising the moulded (thermoplastic) casein (amino acid polymer) to cause cross linkage.

Phenolic plastics are thermosetting, made by interacting a phenol with an aldehyde; for example, phenol formaldehyde (bakelite) and resorcinol formaldehyde types (PF and RF respectively).

Amino plastics include urea formaldehyde (UF) and melamine formaldehyde (MF) types. They are thermosetting.

Polyvinyl chloride (PVC) and polyvinyl acetate (PVA) are vinyl plastics made from the monomers vinyl chloride and vinyl acetate, prepared using ethylene. They are thermoplastic. PVC may be plasticised (rubbery) or unplasticised (rigid).

Polystrene (polyvinyl benzene) is a thermoplastic, naturally transparent and brittle, but it can be toughened in manufacture. It is important in its expanded form.

Polythene (polyethylene) and polypropylene are similar thermoplastics made by compressing the heated gases ethylene and propylene respectively. Polythene is made in 'medium' and 'high' densities, often carbon-pigmented to resist sunlight. Polypropylene has a higher softening temperature and greater resilience than polythene.

Polymethyl methacrylate ('Perspex') is a thermoplastic acrylic resin, useful for its high natural transparency.

Nylons are thermoplastic polyamides which can have high softening temperatures, strength and toughness and are self-lubricating.

Polyesters, made by reacting certain organic acids and alcohols, can be thermosetting or thermoplastic. Those used in paints are called alkyd resins.

Epoxy resins and polyurethanes are used mainly for their toughness, flexibility and adhesion in paints and coatings.

Synthetic rubbers include the polymers polyisoprene, polychloroprene ('Neoprene'), polybutadiene ('Buna') and the copolymers acrylonitrile-butadiene (nitrile rubber), iso-butylene-butadiene (butyl rubber) styrene-butadiene and ethylene-propylene.

Silicones are polymeric compounds based on silicon and oxygen which have high water-repellency and resist extremes of temperature.

Plastics are tested for colour fastness, dimensional change and loss of volatile matter on heating, curling, indentation, deflection and impact resistance, and resistance to chemicals, oils and organic solvents.

14

ADHESIVES AND MASTICS

Adhesives and mastics are related products because of the frequent need for a mastic to have adhesive properties and the use of synthetic polymeric compounds in both groups.

ADHESIVES

The development of adhesives has now reached an advanced stage. There is ample choice of available types and grades to meet all user requirements.

The use of adhesives for the jointing of structural members, especially in timber engineering, is proof of the strength and durability which may be achieved with modern adhesives. However, you must always see that the adhesive you use is suitable for the particular purpose, and ensure that the conditions and method of application are appropriate to the grade of the adhesive used. The instructions of the manufacturer should be carefully studied, since methods of application, curing times and performance vary according to formulation, and different grades of the same name product are often manufactured to meet different user requirements. The intermixing of different types in a product is also a feature; for example among the synthetic resin and solvent adhesives. Another factor is the large number of proprietary or trade names applied to products of any one class, although you will often be able to differentiate between them by their description, trade literature and performance.

Bond

Bond is obtained most readily by *mechanical adhesion*; for example, on keyed, roughened or porous surfaces.

For smooth, impervious surfaces without mechanical key only *specific adhesion* (inter-molecular attraction) can be relied upon. In these cases, it is essential to have a chemically clean surface and a suitable type of adhesive. An adhesive with a chemical set, a tack-drying solvent or a hot-poured mastic type would be best.

TYPES OF ADHESIVE

Animal glues

Animal glues are made from the extract from bones and hides. They are caused to set or harden either by cooling or by removal of moisture from the adhesive, or a combination of both. There is no chemical change involved. Moisture is usually removed by absorption into the material after it has been glued, or by evaporation. Removing the moisture or cooling the hot glue reduces the mobility of the liquid until it becomes rigid. The loss of moisture is accompanied by shrinkage. The whole process is reversible, so the hardened glue will soften and swell if moistened or if kept in air at high humidities, its strength being reduced accordingly. This makes such glues unsuitable for use externally or in damp conditions, unless measures are taken to exclude the damp. If damp, these glues are also liable to destruction by micro-organisms (moulds and bacteria). However, it is sometimes an advantage to use a glue which can be temporarily re-softened at an intermediate stage in the manufacture. This is one advantage of these glues.

Animal glues are conveniently applied without special control measures and are relatively inexpensive. They are supplied in gel form for hot application, in liquid form for cold application and in powder, granular or cake form, for water addition and heating as required. They are particularly useful for woodwork.

BS 745 'Animal glue for wood' deals with the general requirements and testing of animal glues.

Casein glues

Casein is a powder derived from milk curd and made water-soluble by the addition of an alkali. The casein glue powder is simply mixed with water for use, and sets partly by loss of water and partly by chemical action. The fact that chemical action is involved means that the glue must be used within a certain period of mixing, known as the *pot life*. This varies from several hours upwards and is reduced at higher temperatures. Joints are usually held under pressure during the setting period.

Casein glues resist moisture to a reasonable extent, but cannot withstand prolonged direct exposure to weathering and are therefore unsuitable for unprotected external use. When damp they are also liable to attack by micro-organisms. Some types have a staining action on wood, with which they are commonly used, but 'non-staining' grades are obtainable.

BS 1444 'Cold-setting casein glue for wood' deals with general requirements and tests.

Synthetic resin adhesives

There are a number of different types of adhesive within this class. They set chemically but may be either thermosetting or thermoplastic. The thermosetting types undergo an irreversible set whereas thermoplastic types can be re-softened by heat and will reset on cooling.

The thermosetting group includes the *amino resins*—urea formaldehyde (UF) and melamine formaldehyde (MF), the *phenolic resins* —phenol formaldehyde (PF) and resorcinol formaldehyde (RF), and the *epoxy resins*.

The PVA glues, based on *polyvinyl acetate*, are thermoplastic types.

Amino and phenolic resins

The setting action of amino and phenolic resins depends on the chemical action of polymerisation which was explained in Chapter 13. These adhesives are not water-soluble. In general, they are highly resistant to moisture and can withstand exposure, but the different types are not equally efficient in these respects. These factors also depend on formulation and conditions of use and curing.

Amino and phenolic resins are supplied either in powder or viscous liquid form, and may or may not have to be used in conjunction with a hardener. Special glue film is available for use in the manufacture of plywood and other laminated products.

A hardener may be mixed with the adhesive before application, or the two may be applied separately to the components to be joined. and then brought together. The setting time can sometimes be adjusted by varying the amount of hardener added, and is also greatly accelerated by the application of heat. In viscous liquid form they cannot be stored indefinitely, since hardening will eventually occur without any hardener being added, usually after many months or even years, depending on temperature.

Where the surfaces to be joined are not in very close contact during setting (i.e. more than 0·125 mm apart), a *gap-filling* grade of adhesive should be used; otherwise there is a tendency for crazing of the *glue line* (the film of hardened glue). This applies to most cases of the jointing of timber. Gap-filling types are specially blended with a plasticiser or filler.

These adhesives may possess acidic, alkaline or neutral properties according to their basic formulation and the nature of the hardener.

BS 1203 and BS 1204 deal with synthetic resin adhesives for plywood and for constructional work in wood respectively. They also specify types of adhesive and give methods of test.

PVA glues

PVA glues are PVA resin, emulsified in water. The setting action is not chemical so the glue is supplied ready mixed and has indefinite storage and pot life if prevented from drying out. Water may be added if thinning is necessary. These features make PVA glues very convenient to use. They re-soften if exposed to moisture and are used mainly for internal work. Relevant British Standards include BS 4071 'Specification for polyvinyl acetate (PVA) emulsion adhesives for wood' and BS 3544 'Methods of test for PVA adhesives for wood'.

Other types of glues

Fish glues, blood-albumen and soya glues are little used in the building industry of Great Britain, but they may be encountered in imported plywoods. Fish glues have properties similar to bone glues, and blood albumen and soya are comparable to casein types. Blood albumen is sometimes mixed with casein.

Starch and flour pastes, supplied in powder or paste form, are prepared simply by mixing with water, for cold application. They require an absorbent surface to assist the drying and are limited to materials such as wallpapers and fabrics because they are water soluble. In damp conditions they can support mould growths, but toxic forms are available. Cellulosic types are used for similar purposes.

Solvent type adhesives include adhesives based on such materials as bitumen, synthetic or natural rubbers, resins or gums, which harden by evaporation of their volatile solvent. Bitumen-rubber compositions are common. Solvent types are useful on non-absorbent materials, which must be dry to achieve their full adhesion. 'Contact' types, usually based on rubber, are allowed to become 'tack-dry' before mating the surfaces. The same base materials are also used as emulsions in water (e.g. rubber latex and bitumen emulsion adhesives) and these dry by evaporation of the water. Emulsion types are suitable for porous materials, and are not limited to dry surfaces.

Bitumen, pitch and tar are also used naturally, as hot-poured adhesives, and become hard as they cool.

Silicate adhesives, based on sodium or potassium silicate, are used for joining glass to glass, or glass to metals.

Precautions

Many glues have to be used with great care because of their alkaline or acid nature. Alkaline types can cause staining of certain timbers (e.g. oak), which may, or may not, be of importance to the work. Corrosion of metals can occur. Users may also face the risk of dermatitis (skin afflictions), and appropriate precautions should be taken.

TESTS ON ADHESIVES

The tests most commonly made on adhesives are tests for dry strength, and wet strength after immersion in cold water, hot or boiling water or steam. A mycological test is sometimes required.

For convenience, all these tests can be made with the same form of prepared specimen for a given adhesive, and the results assessed by a strength test. For the mycological test the general procedure given in Experiment 31 is suitable, followed by a strength test. Dry strength tests are sometimes made on specimens cured at various temperatures, and after different curing periods. They can also be made to determine ageing effects. For the wet strength tests the method and period of conditioning relate to the type of adhesive, or specification requirement.

A form of test piece made with European beech slips used for adhesives is shown in Fig. 69a. After gluing, the single lap joint is held under pressure in a simple clamp designed to apply a load of 45 kgf to it until the adhesive is set or cured. After conditioning as required the specimen is tested by gripping its ends in the jaws of a tensile testing machine and applying a load to failure. The rate of loading may be specified.

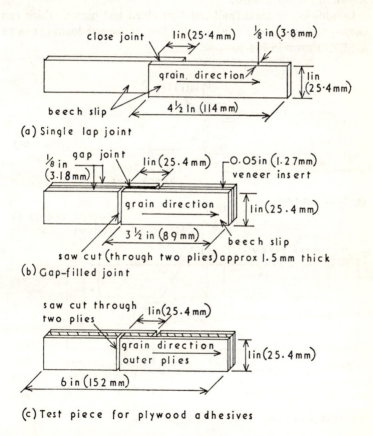

(a) Single lap joint

(b) Gap-filled joint

(c) Test piece for plywood adhesives

Fig 69 Test pieces for strength tests on adhesives ('Pull tests')

When testing a gap-filling adhesive the modified form of test piece shown in Fig 69b is used. This requires two full length beech slips and two shorter veneer inserts. One slip is first coated over approximately the middle two-thirds of its length with adhesive, the two inserts are fitted as shown, and the resultant gap is filled with an excess of adhesive. A second slip is coated as before and superimposed on the inserts. After curing in a clamp screwed down 'finger-tight', saw cuts are made as shown and the specimen is ready for test.

Where the adhesive is specifically for plywood, the test piece shown in Fig 69c is used.

In addition to these 'pull tests' on glued test pieces, which can cause problems due to non-axial loading, transverse loading can be used, as shown in Fig 70.

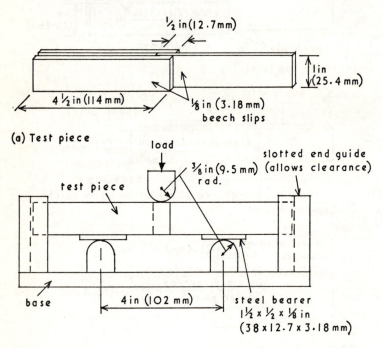

$\frac{1}{2}$ in (12.7mm)

1 in (25.4 mm)

$4\frac{1}{2}$ in (114 mm)

$\frac{1}{8}$ in (3.18 mm) beech slips

(a) Test piece

load

$\frac{3}{8}$ in (9.5 mm) rad.

slotted end guide (allows clearance)

test piece

base

4 in (102 mm)

steel bearer
$1\frac{1}{2}$ x $\frac{1}{2}$ x $\frac{1}{8}$ in
(38 x 12.7 x 3.18 mm)

(b) Test method

Fig 70 Transverse test on adhesives

Mastics are materials used to seal gaps or cracks, or to act as a joint filler or a jointing agent. They remain plastic or flexible to a sufficient degree to permit relative movement between the surfaces, sealed or jointed. It is essential that they are able to adhere firmly to the surfaces to prevent unsealing or loss of filler. In addition, they should not crack, run, blister or exude under normal temperature variations. Mastics for external use must in addition be weather resistant. The choice of a suitable mastic must also take account of special requirements such as resistance to chemicals (e.g. alkalis or acids) and solvents (e.g. petroleum and oils). When used as a jointing medium—for example, to fix tiles or glass—they are also acting as adhesives.

Materials used as mastics include *plastic* materials such as bitumen and tar, some natural and synthetic resins, as well as *flexible* (resilient) types, which include compositions based on oils, natural and synthetic resins. The flexible types are also referred to as *elastomers*.

Mastics may be applied by hand using a knife or trowel, extruded from a tube or pressure-gun, or applied in the form of tape or strip. Some may be heated and hot-poured. Some stiffen after application by loss of solvent or by chemical action. Others harden at the surface only, to form a protective skin. To obtain good adhesion, surfaces should be clean and dry and in some cases a special priming coat must be applied.

Many *window putties* harden and so are not true mastics. Glazing putties for wood frames are usually a mixture of linseed oil and whiting (powdered chalk), sometimes together with white lead. However, exceptions occur in the case of some glazing putties for metal frames which harden at the surface only, to form a skin enclosing a soft interior.

Summary

Adhesives should be selected according to type and grade to meet the particular requirements of the work.

Types include animal and fish glues, casein and blood albumen, synthetic resins, starch and flour paste, cellulosic types, solvents and emulsions.

Animal glues, from bones and hides, harden by gelling and are re-softened by moisture, but are suitable as an interior adhesive. Fish glues are similar.

Casein glues, derived from milk, set partly by loss of water and partly by chemical action. They are reasonably moisture resistant but require protection if used externally. Blood albumen and soya glues have similar properties.

Synthetic resin types may be thermosetting or thermoplastic.

The amino resins (UF, MF) the phenolics (PF, RF) and epoxies are all thermosetting, have high resistance to moisture and are available for external use.

PVA glues are thermoplastic synthetic resin (polyvinyl acetate) emulsified in water. They are re-softened by moisture but are popular for internal uses.

Bitumen, rubber and synthetic rubber based adhesives are available both in solvent and emulsion forms. Solvent types require dry surfaces. Emulsions can be used on damp (or dry) surfaces.

Bitumen, pitch and tar are also used as hot-poured adhesives.

Silicate adhesives are used for jointing glass.

Strength tests are made on adhesives, using beech test pieces, for dry strength and for wet strength (after immersion in cold, hot or boiling water or steam) to assess their moisture and weather resistance, and at the conclusion of mycological tests.

Mastics are of two kinds—plastic materials, and flexible materials (elastomers). Both are required to have good adhesive power and should neither slump nor crack.

They include oils, tar, bitumen, natural and synthetic resins.

15

PAINTERS' MATERIALS

Painting is the application of a thin coating of material in a liquid or plastic condition to a surface, as a decorative or protective measure. The applied coating is normally required to dry out or harden to a solid film. The coating applied can vary widely in its composition and chemical and physical properties, depending on the nature of the surface being coated, the method of application and the requirements in regard to decoration and protection for the work in hand. The materials used by the painter are of different classes, such as paints, varnishes, stains and preservatives (see Chapter 11). The paint may be applied by brushing, spraying, dipping, or by roller. In the early days of the craft the painter mixed his own materials, and the exact ingredients and their proportions were regarded to a certain extent as secrets of the trade. Eventually, and inevitably, there appeared on the market the *ready-mixed* products, now made by specialist manufacturers to proprietary formulations, or to the larger customers' specifications.

The range of natural and synthetic (man-made) materials now being used in the making of paints, and the range of different products now available, is so vast that a simple classification of these would be very difficult. It is therefore important that the user should take careful note of the manufacturer's instructions and recommendations regarding selection, storage, preparation, application, subsequent treatment and maintenance of their products. Their advice should be sought when in doubt.

PAINT COMPOSITION

A paint consists essentially of one or more solid materials—the pigment, and a liquid known as the *vehicle* or medium.

Pigments

The pigment is usually present in powder form, in suspension in the vehicle. Its function may be to provide the opacity ('hiding power') and colour of the paint film, to confer durability or corrosion resistance to the surface painted, or both. Pigments are, in the main, inert white or coloured powders, such as metallic oxides or salts, although they may in some cases have a chemical influence on the hardening of the paint film. They are obtainable in powder form, or sometimes mixed with oil or water as a paste.

Among the natural white pigments are barytes (barium sulphate), whiting or Paris white (chalk), China clay or kaolin, and terra alba or gypsum (calcium sulphate). Artificial white pigments include white lead (carbonate and hydroxide of lead), titanium oxide, lithopone (a calcined blend of zinc sulphide and barium sulphate), zinc white (zinc oxide) and antimony white (antimony oxide). In general, the artificial whites mentioned give better obliteration (opacity) than the natural whites, but this can depend on the particular vehicle with which they are mixed. Pigments of low obliterating power are frequently added to high opacity pigments because they impart some other desirable property; for example, they make application easier, or may assist the suspension of the solid particles, in which case they are referred to as an extender.

Black pigments include carbon black, lamp black and vegetable black, all forms of carbon obtained by burning organic matter in a restricted supply of air or oxygen.

Among the colour pigments ochre and sienna (yellows), umber (brown), venetian red and red ochre are examples of earth pigments, i.e. occurring naturally in the earth's crust. They owe their colour to the presence of iron compounds. Some earth pigments are also made artificially; for example, ultramarine blue. Other pigments are the products, or by-products, of chemical processes; for example, the well-known and important Prussian blue. Lake pigments are a group of pigments made by impregnating powder materials with organic dyes (e.g. crimson lake and rose pink).

The adjustment of colour in a particular paint may involve the mixing of one or more colour pigments together with a white or whites.

Pigments specially valuable for their corrosion-inhibiting properties are red lead (an oxide of lead used for painting iron and steel), zinc chromate (for aluminium, iron and steel) and calcium plumbate (for zinc, galvanised iron, iron and steel), in addition to the metallic powders, lead, zinc and aluminium.

Paint vehicle

The vehicle, or medium, is the liquid part of a paint. Its primary function is to facilitate application by giving the paint mobility (the ability to flow), but it often serves also as a binder for the pigment on drying, and gives adhesion to the surface painted. In fact, the nature of the vehicle will largely determine the characteristics and properties of the hardened (dry) paint film. The vehicle may also include a thinner, such as turpentine (a tree extract) or white spirit (a petroleum derivative), added to give increased mobility. Water, various volatile solvents (often with gum or resin content), emulsions, drying oils and oil varnish are used as vehicles.

A volatile solvent is a liquid which dissolves a substance (the film-forming constituent) but which readily evaporates, leaving behind the dissolved substance unaltered chemically. These include various alcohols (e.g. methylated spirit), naphthas and other coal-tar derivatives, and a very wide range of other organic chemicals.

A *gum* is a carbohydrate soluble in water, an example of which is gum arabic, a tree extract.

A *resin* is either a natural extraction from certain trees (e.g. copal) or insects (e.g. shellac), or a material of similar character of synthetic origin. They are polymeric substances and unlike gums, resins are not soluble in water, but are soluble in certain other solvents (see Chapter 13).

A *drying oil* is an oil (a natural extract, or synthetic) which has the property of hardening to a paint film when exposed to the atmosphere, wholly or partly by chemical action with absorbed oxygen. Drying oils may be natural extract or synthetic and are mainly vegetable extracts, such as linseed oil from the seed of the flax plant, tung oil from the seeds of certain trees, and soya bean oil. Such oils are usually subjected to special processing in order to reduce their natural drying (hardening) time, which is often excessive. For example, linseed oil has a natural drying time of many days or weeks, but when subjected to certain heat treatments (e.g. as in the case of 'blown' linseed oil, 'stand' oil and 'boiled' linseed oil) this is considerably reduced.

Oil varnish is a substantially transparent, pigment-free composition of drying oils, natural or synthetic resins and solvents. These compositions are also known as oleo-resins.

Other paint ingredients

Other *special ingredients* which may be present in small amounts include *driers* to assist hardening, *plasticisers* to improve the flexibility of the hardened paint film and, as previously mentioned, *extenders* to improve working characteristics and film-forming properties, or to assist the suspension of the pigment.

PAINT TYPES

The main types of paint are emulsion paints, water paints and distempers, solvent paints, oil paints, hard gloss paints, enamels and synthetic resin paints.

Emulsion paints

The typical emulsion paint is a synthetic resin (e.g. polyvinyl acetate —PVA) emulsified in water, but other emulsifiable binders are oils, oil varnishes, resins, rubbers and bitumen. Other ingredients are usually stabilisers, which prevent coagulation. The drying action of these paints is due to evaporation of the emulsifying liquid. A useful feature of these paints is that their paint films are initially, and sometimes permanently, permeable, which makes some suitable for decorating new plaster and other damp surfaces, since background moisture can dry out through the paint film. Their hardened films are washable, which improves with age, and some are suitable for external use. PVA emulsion paints are alkali-resistant and so may be used on cement, concrete, asbestos-cement and plaster. They are suited to roller and brush application.

Water paints

Water paints have a vehicle composed of a drying oil, oil varnish or synthetic or natural resin emulsified in water, usually together with a stabiliser such as glue size or casein. Water paints are, in fact, emulsion paints, although they are not classified as such commercially. Pigments and extenders are added and the product is usually supplied in the paste form mixed with water, to be thinned with water by the user to the consistency required for application. They give a permeable paint film which is washable when hard, and most are unaffected by alkalis. They are used mainly for interior decoration, but some are suitable for external use.

Distempers

Distempers differ from water paints in that they do not contain a drying oil, oil varnish or resin and they are not emulsified. They consist of a pigment and extender with a water soluble binder such as glue size, and are supplied either in the mixed powder form or as a paste in water. They are prepared for use by adding water. As distempers are suitable for internal use only they are also known as soft distempers to stress the fact that they are non-washable and easily rubbed off. Their main use is on ceilings, or as a temporary decoration for new walls.

Solvent-type paints

These include various forms of rubber-based paint, such as cyclised rubber and chlorinated rubber varieties, bituminous and tar paints and cellulose enamels. These paints dry essentially by evaporation of the volatile solvent and care is needed in their use because their vapours are often inflammable and may also be toxic to the user if inhaled. Two or more separate applications of these paints are usually needed to build up the required finished thickness, and care is needed to avoid 'lifting' the previous coat due to the solvent action of a later application.

Both rubber-based and bituminous paints are particularly resistant to water, water-vapour and most chemicals, but should only be applied on dry surfaces. Most are subject to gradual degradation when exposed to direct sunlight, resulting in colour-fading (or greying in the case of blacks), and to progressive reduction of film thickness due to weathering. They have been widely used on both metals and alkaline surfaces (concrete, rendering and asbestos cement).

The use of cellulose paints, based on nitrocellulose, is confined mainly to spray application on metals and is characterised by the high gloss and wide range of colours obtainable, but with comparatively thin finished film thickness.

Oil paints

These have a vehicle consisting of a drying oil, or oil varnish, mixed with a thinner. They are typified by the group based on linseed oil with white spirit, which constitute the traditional ready mixed oil-based paints, but other natural and synthetic oils and thinners are also used. These paints are obtainable in a wide variety of colours and suit most purposes for both internal and external use.

Hard gloss paints

These have a vehicle consisting of a specially treated oil varnish or drying oil (with or without resin) mixed with a thinner. They are capable of giving a better gloss than ordinary oil paints, and are often more rapid in drying. They can be used externally, although their durability may not equal that of a good quality oil paint.

Enamel paints

An enamel is a variety of hard gloss paint with the special property of drying to an exceptionally brilliant gloss finish, and possessing remarkably good flow characteristics in application. This is usually achieved by using best quality pale elastic (e.g. copal) varnish and limiting the amount of pigment added. They are not usually suited to external use, as they are less durable to weathering than a good quality oil paint.

Synthetic resin paints

These have as their vehicle a synthetic resin, with or without drying oils and solvents.

A good example is provided by the alkyd paints, which are paints based on alkyd resins, often with added drying oils, which can offer high gloss coupled with long life under severe exposure. These also have good flow properties in application and are quick-drying, with good adhesion and flexibility.

Other types are the epoxy and urethane paints based on epoxy and polyurethane resin respectively. Both are obtainable as *two-pack* products, in which case one pack contains a curing agent which sets off the hardening action as soon as they are mixed together. Poly-urethane types are also available as a one-pack product. These paints are obtainable coloured, or clear for use as varnishes. They are highly resistant to water and chemicals and are capable of providing a very hard-wearing surface. Uses include their application to timber, concrete and metals.

SPECIAL PAINTS

Silicate paints

These are based on water-soluble alkaline silicates (e.g. sodium silicate), which become insoluble on drying. Their main uses are as a weather-protective on concrete, brick, asbestos-cement and similar surfaces, and as fire-retardants.

Neoprene paints

These are based on a proprietary synthetic rubber. They are usually a two-part type with a vulcanising agent as 'catalyst'. They are chemical and weather resistant.

Cement paints

These non-conventional type paints consist essentially of a white or coloured Portland cement in powder form, to which water is added. They are used mainly on concrete and rendering.

Plastic paints

This name is given to paints which are applied to the consistency of a thick cream, which allows them to be textured before they set or harden. They are based on such materials as gypsum, or whiting, with a gelatinous binding agent.

Thixotropic paints

These are paints which have the character of a thick cream or gel when undisturbed, but which are free-flowing when subjected to a shearing action as during brush application. They are also referred to as 'one coat', 'non-drip' or 'non-sag' paints.

Stoving paints or enamels

These are paints designed to be cured (hardened) by baking, usually at a temperature above 60°C, after application, for example, in an oven or by exposure to infra-red radiation. Included in this group are varieties of the thermosetting synthetic resin types.

Silicones

These include a full range of synthetic paint types, based on silicon-oxygen compounds, including water-repellent solutions and emulsions, heat-resisting resin-based paints, and silicone-organic copolymer types. Silicon liquids are also used as additives to organic synthetic resin paints as modifiers.

OTHER FINISHES

Stains

These are used mainly to colour or tint new timber, without obscuring its natural grain or figure. They consist of dyes or translucent pigments in water, an organic solvent or a drying oil, sometimes with a preservative. Some are suitable for external use alone and can give protection equal to paint treatments; others can be overpainted with a clear varnish.

Oil varnish

The use of oil varnish as a paint vehicle has already been mentioned. It is also used itself as a clear, or tinted finish, mainly on timber. For external use a suitable grade is required, and up to four coats may be specified for some types to give good durability.

Spirit varnishes

These consist of a resin dissolved in an organic solvent such as alcohol or turpentine.

French polish

This is a spirit varnish based on shellac and methylated spirit. It is applied as an interior finish on timber, using specialist techniques to obtain a lustrous, translucent film.

Wax polishes

These are based on various natural and synthetic resins and are used mainly on interior timber surfaces, such as floors, to maintain gloss and protection against wear. Emulsified varieties are available.

MISCELLANEOUS PRODUCTS

Knotting

This is a sealer applied to the knots of resinous timber before painting, consisting usually of shellac dissolved in methylated spirit.

Size

This is a solution of a water-soluble glue of animal extract, starch or cellulosic types.

Clearcole

This is a prepared solution of size tinted with whiting, used to adjust the suction of a plastered surface before applying a size-bound distemper.

Petrifying liquid

This is water-paint medium, used either as a thinner for oil-bound water paint or to precede its application to adjust suction.

Driers

These are substances added to paints to reduce their drying time; for example, compounds of lead, manganese or cobalt in the case of oil paints. For convenience they are usually blended with oils and solvents in a liquid preparation.

PAINT SCHEMES

Some paint treatments consist simply of applying one or more coats of a single material, possibly varying the amount of thinners in successive coats. Examples include the solvent-type paints, and emulsion paints. In other cases successive coats may need to be of different composition, in which case the whole process is referred to as a paint scheme. The need for a paint scheme arises from different requirements in a paint finish, such as gloss, opacity, hardness,

permeability (or impermeability), corrosion-inhibiting properties, chemical resistance, adhesion and compatibility with a given surface or a preceding coat. Not all of these factors will usually be obtainable in the required degree from a paint of one particular composition, and indeed some may conflict. For example, with an oil paint an increase in the ratio of pigment to oil will lead to greater opacity but with corresponding reductions in gloss and, usually, durability. In practice, a full paint scheme using one of the traditional finishes such as oil paint, hard gloss or enamel will normally require at least three coats, each of different composition and referred to respectively as the priming coat, the undercoat and the finishing or final coat.

Primer

The function of the priming paint or *primer* may be to give protection against corrosion of metals or against dampness, especially of site-stored joinery. Priming paint can also be used to adjust the suction of a surface (e.g. of plaster, concrete or timber) and provide good adhesion for subsequent coats. Alternatively, it can act as a *barrier* coat to isolate one coat from a preceding coat; for example, to prevent chemical interaction.

Mention has already been made of the main corrosion-inhibiting pigments, and these form the basis of well-known priming paint types such as *red lead* primer based on red oxide of lead Pb_3O_4 and used on iron and steel; *calcium plumbate* primer for zinc and galvanised iron (without preliminary etching) and steel; *zinc chromate* primer for aluminium, *metallic lead* and *zinc-rich* primer ('cold galvanising') for iron and steel and *metallic aluminium* on copper or brass. Another primer used for iron and steel, and metals generally, is *red oxide* primer (based on red oxide of iron, Fe_2O_3), although this is regarded as less effective than red lead in protecting ferrous metals.

A common primer for timber is one based on white lead mixed with a smaller amount of red lead, often referred to as *pink* primer owing to its colour; white lead is lead hydrocarbonate, $2Pb\,CO_3.$ $Pb(OH)_2$. This has the advantage that it is compatible with most other paints likely to be used in subsequent coats. Aluminium primer (based on the metal) is particularly useful as a sealing coat, as it will act as a barrier to vapour, to resinous knots in timber and to bitumen. Resins and bitumen will soften most other paints and 'bleed' through them. The pigment in aluminium paints is either in leaf or powder form, giving paints described as *leafing* and *non-leafing*. The leafing type is used for priming since it has the better sealing action, whereas the non-leafing type gives better lustre and is preferred as a finish.

An *alkali-resistant* primer is one designed to be used as a barrier coat on alkaline surfaces, such as cement or lime plaster and concrete, where the subsequent paint coats are not themselves resistant to alkalis.

Undercoat paints

The function of an undercoat is mainly to obliterate the background and provide a uniformly dense tone or colour to assist the finish. It is also required to provide good adhesion for the finishing coat. It follows that undercoat paints tend to be heavily pigmented and usually need to be sealed against the weather by a finishing coat.

Finishing paints

These give the required reflection characteristics to the painted surface (flat, eggshell or high gloss) and of course give the final colour. In the case of exterior paints, finishing paints must also seal the surface against the weather, and some elasticity may be desirable.

PAINT FILM DEFECTS

Defects caused by dampness

Any paint film applied to damp materials and which is not appreciably permeable will be likely to fail by softening, blistering, peeling or flaking.

If the background is of an alkaline nature the presence of moisture can cause bleaching of certain pigments and softening of an oil-bound film due to chemical action, with an eventual complete breakdown. This softening action is known as alkali-attack, or saponification.

Where soluble salts are present in background materials, dampness can lead to their crystallisation beneath a paint film which is not substantially permeable, and can cause unsightly efflorescence on the surface of permeable films.

I

Mould growths, often with the appearance of stains, can occur in damp conditions where a food source is available, such as carbohydrates (e.g. in size-bound distempers, or sapwood). These can be controlled by the use of antiseptic solutions, but the source of the moisture must be found and removed.

Crazing or cracking

This takes various forms in paint films and may either be the normal form of deterioration due to weathering, or a symptom of unsuitable paint formulation, adulteration, wrong selection or application techniques, or subsequent maltreatment.

Chalking

This is the powdering of a paint film due to normal weathering, or to adverse conditions, causing breakdown of the binder.

Bleeding

This results from the softening of a paint film due to an agency such as resinous or bituminous matter acting as a solvent on the binder. The solvent gradually penetrates the paint film and causes staining and local breakdown. A common example is the breakdown of an oil or resin paint film on timber due to resin exuded from knots. Bleeding can be prevented by the use of a suitable sealer as a barrier coat; for example, shellac 'knotting' applied to knots in timber, or an application of leafing aluminium paint on bitumen-coated surfaces.

Sulphiding

This is the discolouration of certain paint films due to the conversion of compounds present to the sulphide; for example, the blackening of lead-based paints due to lead sulphide formed by the action of sulphurous fumes in an industrial atmosphere.

STORAGE OF PAINTS

Since most paints are essentially mixtures, their constituents are liable to separate out to form different layers in the container during storage. Thorough remixing should therefore be the rule before and during use. Whenever a skin forms at the surface, it will usually re-dissolve on mixing in solvent paints, but should be removed with non-solvent paints by straining the paint through a fine-mesh fabric.

Some paints need protection from excessive cold; for example, water paints and emulsion paints are likely to deteriorate at near-freezing temperatures.

Paints containing organic solvents present a fire-risk, and many can give off toxic vapours, so the necessary precautions should be taken.

BS REQUIREMENTS FOR PAINTS

The principal current British Standards for paints are the series dealing with ready mixed oleo-resin based paints (BS 2521–32), and BS 1053 for water paints and distempers. In addition to this, there is BS 1391 which specifies tests for anti-corrosive protective schemes applied to steel and wrought iron and BS 2660 which gives a system of classifying 'standard' colours by serial number references. An outline of the tests and requirements of some of these British Standards will give an idea of factors significant to the paint technologist.

BS 2521–32

These specify the composition of ready mixed oleo-resin based paints, giving pigment: binder: solvent ratios. Also specified are consistency (by verbal definition or rate of flow measured using a standard flow cup); drying time (for the surface dry and recoating conditions); finish (degree of gloss); water content (maximum) keeping properties (storage life); and, for undercoat and finishing paints only, colour (to BS classification); opacity, which is determined by the light transmitted through a paint coat applied on a transparent cellulose film measured using a reflectometer. The apparatus consists of a lamp source, a photo-electric cell and a sensitive galvonometer. The result is expressed as a 'contrast ratio'. Methods of test are given.

BS 1053 'Specifications for water paint and distemper for interior use'

This distinguishes water paints (washable, oil-bound) from distempers (non-washable oil-free) by specifying their general composition and washability characteristics. The following tests are described, and requirements are specified for results: consistency (by verbal definition), colour and finish (by film characteristics and colour-match with sample or other agreed standard), reflectance value of white paints only, as percentage of light reflected normally from the surface, measured using special apparatus comprising a photo-electric reflectometer and sensitive galvanometer, content of non-volatile matter, resistance to dry rubbing (test on a dry film) and keeping properties (storage life). In addition, for oil-bound washable water paints, there are tests for oil/resin and pigment content, extracted by solvent and chemically analysed, and recoating to see if a brush application will unduly soften and lift a previous coat.

BS 1391: '*Performance tests for protective schemes used in the protection of light-gauge steel and wrought iron against corrosion*'

Details are given of two tests applied to assess the performance of paint schemes:

[*a*] A.R.E. *salt droplet test*—suitable for testing both prepared panels and actual articles. The apparatus is easily improvised. The test area is periodically sprayed with a salt solution of specified composition (simulating sea-water), while stored in air at high relative humidity. The paint scheme is assessed on its effectiveness in preventing corrosion, and limits are given for the extent of rust and rust staining in a standard test. The test also allows a direct comparison of different paint schemes tested simultaneously.

[*b*] C.R.L. *sulphur dioxide test*—suitable only for testing prepared panels. Special apparatus is required, comprising an enclosed beaker with both a heating and a cooling device to cause condensation on the test surfaces. Results are assessed in the same way as for the A.R.E. test.

PAINT TESTING

A full range of equipment is available for the testing of paints to commercial standards. The properties measured include particle size (pigments), viscosity, weight per unit volume, film thickness (wet and dry), hardness, adhesion and flexibility, drying time, opacity, colour, gloss, resistance to *abrasion* and artificial weathering.

Artificial weathering is an accelerated test in which painted specimens are subjected to various agencies such as heat, light and ultra-violet radiation, with alternate periods of water-spray and drying, and grit-blasting. The results of such tests are useful but require careful interpretation, and long-term tests with specimens exposed to natural weathering are more reliable. In these *exposure tests*, painted metal or timber panels (usually 300 × 150 mm) are set in a frame at 45° to the horizontal and facing south. They are inspected at intervals, usually over a number of years, to assess any deterioration of the paint scheme.

EXPERIMENT 32 Specific gravity and weight per unit volume of paints

Note Standard density cups are obtainable and could replace the measuring cylinders used in this experiment

Apparatus Measuring cylinders (100 ml)—beakers (150 ml)—mixing knife or spatula—drying cloth and cylinder-cleaning brush—physical balance (to 0·1g)

Specimens Ready-mixed paints (oil, hard-gloss, enamel or emulsion); some white spirit or other appropriate paint solvent

Method (for each specimen)
1 Weigh the measuring cylinder empty
2 Stir the paint to mix it thoroughly and fill a beaker
3 Pour the paint into the measuring cylinder up to the 100 ml mark
4 Weigh the filled measuring cylinder
5 Empty the cylinder and clean it with solvent

Results Set these out as follows, for calculation:

Weight of filled cylinder (A) = gf
Weight of empty cylinder (B) = gf
Weight of 100 ml of paint (A−B) = gf (C)
Specific gravity (C÷100) =

This is numerically equal to the weight per unit volume in *kilograms—force per litre.*

Alternatively weight per imperial gallon (s.g. × 10)=lbf/gal

Example 24 A measuring cylinder weighs 90·5 gf empty and 280·5 gf when filled with paint to the 100 ml/mark. Calculate [a] the specific gravity and [b] the weight per litre and weight per gallon of the paint.

[a] Weight of 100 ml of paint = 280·5 − 90·5
                                = 190·0 gf
                    s.g.   = 190 ÷ 100
                                = 1·90 *Ans.* (a)
[b] i Weight per litre   = 1·90 kgf/l *Ans.* (b) i
[b] ii Weight per gal (Imperial) = 1·90 × 10
                                = 19·0 lbf/gal *Ans.* (b) ii

EXPERIMENT 33 Spreading power (coverage) of paints

Note Paints used must previously have been tested for specific gravity (see Experiment 32)

Apparatus Plywood panel, area 1 ft² or 1 dm² (0·01 m²)—surface primed ready to receive undercoat and finish paints—beakers—palette knife—physical balance (to 0·1 g)—paint brushes (1 in/25 mm)—drying cloth
Specimens Undercoat and finish paints

Method
Stage 1—Undercoat
1 Remix the paint thoroughly and pour some into a small beaker
2 Weigh the beaker of paint, together with the brush
3 Apply by brush one coat of the undercoat paint to the panel (over 1 ft² or 1 dm² area),* avoiding loss of paint by spillage, etc
4 Replace the used brush on, or in, the beaker of paint and re-weigh them together to determine the amount of paint used

Stage 2—Finish
After allowing the necessary time for the undercoat paint to dry, repeat the procedure of Sections 1 to 4 using the finishing paint

Results Set these out for calculation as follows:

	Undercoat	Finish
Initial weight of beaker + paint + brush (A)		
Final weight of beaker + paint + brush (B)		
Weight of paint applied (A − B)	gf(W)	gf(W)
Specific gravity of paint (G)		

Calculation
* Spreading power

[a] in *square yards per gallon* from

$$4540 \, A \times \frac{G}{W} =$$

where A = test area in yd^2

or [b] in *square metres per litre* from

$$1000 \, A_m \times \frac{G}{W} =$$

where A_m = test area in m^2

Note The result can only be approximate due to the small test area used and will relate to the consistency on application. For greater accuracy use a larger area.

EXPERIMENT 34 Paint film drying times

Apparatus Plywood panel (150 × 100 mm)—surface primed ready to receive undercoat and finish paints—beaker—palette knife—paint brushes (1 in/25 mm)—drying cloth—clock—clean, fine sand (BS 52–100 mesh)

Specimens Undercoat and finish paints
1 Thoroughly mix the undercoat paint and apply a single brush coat to one face of the primed test panel. Note the time at which the application is made
2 Leave the panel to dry in a dust-free room atmosphere, laid horizontally with the painted face upward
3 At intervals (e.g. 5 min) check the condition of the paint film for drying. *Note* The following definitions apply:
[a] *Surface-dry* The paint film is dry on the surface but soft underneath (fine sand ceases to adhere to the surface)
[b] *Touch-dry* Slight finger pressure leaves no mark on the paint film
[c] *Hard-dry* The surface is capable of receiving a further brush-coat. The paint film ceases to be tacky and is not marked by a thumb-nail applied with light pressure. This condition is more readily established using a special test-machine incorporating a loaded needle
4 Following attainment of the hard-dry condition, repeat the procedure of Sections 1 to 3 after applying the finish coat

Results Record the drying times for each paint

Note A result relates only to the conditions (e.g. temperature and relative humidity) under which the test was made

EXPERIMENT 35 To show the permeability of a paint film

Apparatus Three identical tin lids (e.g. 100 mm diameter × 20 mm deep)—plaster of Paris (powder) in normal quick-setting form—mixing board and palette knife—measuring cylinder (100 ml)—beaker—damp closet—paint brush (1 in/25 mm)—physical balance (to 0·1g)

Specimens Water paint or PVA emulsion paint

Preparation Mix up a weighed amount of plaster of Paris with a known quantity of water to produce sufficient paste of a moderately stiff plastic consistency to fill one tin lid (used as a 'dish'). Immediately fill the lid, striking off surplus plaster, and smooth the surface. Fill the remaining lids in turn, using exactly the same amounts of plaster and water as before. Allow the plaster to harden, then place the lids overnight in a humidity cabinet at or near 100% R.H. to prevent the plaster drying out

Method
1 Remove all three plaster-filled lids from the humidity cabinet
2 Apply a first brush-coat of paint to the plaster surface of each of two lids, and leave all three lids in a normal room atmosphere
3 After 4 hours apply a second coat to one of the painted 'lids'
4 After about 20 minutes weigh each of the three 'lids'. The unpainted one will act as a 'control'
5 Weigh the lids periodically (e.g. daily, or at 1, 2, 7, 14 and 28 days)

Results Tabulate the results to show the weight loss by evaporation from the plaster through the paint films, expressed as a percentage of that for the unpainted 'control', and illustrate this by a graph.

Example 25 In a permeability test (see Experiment 35) using two painted specimens and an unpainted control the results were as follows:

Specimen	Weight (gf)	
	Initial	At 7 days
Control	290·0	240·0
1 coat	291·5	250·0
2 coats	292·0	261·5

Express the results for the painted specimens to show the relative loss by evaporation compared with the unpainted specimen.

Calculation:

Specimen	Loss in weight at 7 days	
	Actual gf	of 'control' loss
Control	50	100%
1 coat	41·5	$\frac{41·5}{50} \times 100 = 83\%$
2 coats	30·5	$\frac{30·5}{50} \times 100 = 61\%$

EXPERIMENT 36 To demonstrate alkali attack on an oil paint

Apparatus Mixing board—trowel—wood float—beaker or paint can —paint brush (1 in/25 mm)—two wood or metal moulds for casting mortar panels (size 300 × 300 and 25 mm thick)—galvanised steel tray (size approx. 350 × 350 × 70 mm)—damp closet (at or near 100% R.H.)

Materials Ordinary Portland cement—washed sand, a blend of equal parts BS 25–52 mesh, i.e. passing 25 mesh and retained on 52 mesh and BS 52–100 mesh (Table 6, page 305, gives equivalent metric sizes)—mould oil (with brush)

Specimen Oil paints (undercoat and finish), preferably blue or green

Preparation Cast two mortar slabs, 300 mm square and 25 mm thick, of proportions 1 part cement to 6 parts sand, by weight (grading as specified) with just sufficient water to moisten the dry materials ('earth-moist' consistency).* Bring the mortar to a level surface by means of the float. It should have a porous texture. Cover the slabs with plastics film for 24 hours, then demould them and place one in a damp closet, leaving the other in a normal room atmosphere, each for at least 16 hours. Finally remove the slab from the damp closet and leave both panels in the same normal room atmosphere for two hours. They are then ready for use.

*Note A mix of 1500 g of cement to 4500 g of each of the two sands with water:cement ratio about 0·40 will be sufficient for two slabs.

Method

1 Paint an area 250 × 250 mm square on both panels using the undercoat paint

2 When touch-dry, store one panel in the damp closet for 16 hours and the other under normal room conditions

3 Remove the panel from the cabinet, leave it for 15 minutes to allow condensation to disperse, then apply a finishing coat to *both* panels. This can be limited to an area 150 × 250 mm, to provide both 1- and 2-coat test areas

4 When touch-dry place the *wet* panel on a tray of saturated sand, which must be kept wet during the remainder of the test. Allow the other panel to dry under normal room conditions

5 Inspect the panels after 48 hours and again at intervals over a period of weeks, noting any deterioration of the paint films, such as loss of colour or softening. Examine the films using a hand lens (× 10), probing beneath the surface of any blisters or pock-marks with a pin

Results Record your observations on the paint films. Account for any deterioration and explain any difference in the performance of the wet and dry films.

Summary

Paints are surface coatings applied for protection or decoration.

Most paints are purchased ready-mixed, for application by brushing, spraying, dipping or roller application.

A paint usually consists of a pigment and a vehicle (or medium).

Pigments confer colour and opacity, and may inhibit corrosion.

The vehicle provides the necessary consistency for application and is often the film-forming constituent and binder.

A *gum* is a carbohydrate soluble in water.

A *resin* may be natural or synthetic and is not soluble in water.

A *drying oil* is one which hardens wholly or partly by oxidation.

An *oil varnish* is a clear liquid composition of drying oils and resins.

Extenders assist suspension of pigment, or improve working or film-forming properties. Plasticisers increase film flexibility. Driers accelerate hardening.

Paints

An *emulsion* paint consists of a liquid (or solid) finely and uniformly dispersed in another liquid.

Emulsion paints (of PVA, oils, oil varnish, resins, rubbers and bitumens) dry by evaporation of the emulsifying liquid, and can be applied on damp surfaces.

Water paints are emulsions of a drying oil, oil varnish or resin in water, with a stabiliser such as glue or casein, pigments and extenders, which dry to a washable film.

Distempers are non-washable paints consisting of a pigment and extender with water-soluble binder, in water.

Solvent-type paints include rubbers, bitumens, tar and cellulose enamels, which dry by the evaporation of their volatile solvent.

Oil paints consist of a drying oil, or oil varnish, with pigments and other ingredients such as driers, extenders and thinners.

Hard gloss paints have an oil/resin vehicle and are specially formulated to be quick drying, with high gloss, and are suitable for external use.

Enamel paints are a type of hard gloss paint (based on a pale elastic varnish), rapid drying and mainly for internal use.

Synthetic resin paints and varnishes include the alkyd, epoxy and urethane types.

Special paints are based on sodium silicate, synthetic rubbers, silicones or Portland cements. There are also plastic paints, thixotropic paints and stoving paints.

Other finishes include stains, oil and spirit varnishes and wax polishes.

Miscellaneous products include knotting, size, clearcole, petrifying liquid and driers.

A conventional paint scheme comprises a priming coat followed by one or more undercoats and finishing coats.

A primer may inhibit corrosion, give temporary protection to unfinished timbers, adjust suction, provide adhesion or act as a barrier coat.

An undercoat provides the necessary opacity and colour density.

A finishing coat gives the required degree of gloss, and acts as a weather seal for external paints.

Primers for iron and steel include red lead, calcium plumbate, red oxide and metallic zinc and lead types.

Metallic aluminium paints are used as primers for copper and brass, and as a barrier coat.

For timber, white/red lead primers and aluminium primers are useful. Alkaline surfaces require an alkali-resistant primer unless the paint treatment is one unaffected by alkalis.

Paint film defects

Dampness in background materials can cause softening, blistering, peeling or flaking of impermeable paints, and promotes alkali attack, efflorescence and surface moulds.

Crazing or cracking may be normal deterioration by weathering or due to faulty selection, mixing, application and cleaning techniques.

Chalking of oil paints is due to breakdown of the binder, e.g. by natural weathering.

Bleeding is due to softening of the paint film by solvent action, especially by resin from knots and bituminous materials.

Sulphiding is the blackening of lead compounds in some paints by a sulphurous atmosphere.

Storage

Many paints have limited shelf life. Some require protection from the cold. Others give off inflammable or toxic vapours.

BS requirements and testing

BS for oil paints specify pigment:binder:solvent ratios, consistency, drying time, finish, water content, keeping properties and, for undercoat and finish paints, colour and opacity.

BS requirements for water paints and distempers relate to consistency, colour and finish, reflectance value (white only), non-volatile content, resistance to dry rubbing, keeping properties and, for water paints only, oil/resin and pigment contents and recoating properties.

Tests for protective schemes for ferrous metals include the 'A.R.E. salt droplet' and the 'C.R.L. sulphur dioxide' tests.

General tests on paints are for particle size (pigment), viscosity, weight per unit volume, film thickness, hardness, adhesion and flexibility, drying time, opacity, colour, gloss, abrasion and weathering.

16

BITUMINOUS MATERIALS AND JOINTLESS FLOORING

BITUMENS

Bitumens are mixtures of hydrocarbons, or derived compounds. They may take the form of a gas, liquid or solid, or they may be viscous. However, the ones important in building are those which are viscous or solid at normal temperatures but which soften and flow on heating; they are classified as *asphaltic bitumens*.

Bitumen occurs naturally, or may be distilled from petroleum. It is impermeable, and resistant to both acids and alkalis. It is black or brown, and similar in appearance to the pitch distilled from coal-tar. Unlike coal-tar pitch, however, bitumen is not brittle when hard, and softens only gradually when heated.

Bitumen is used as a binding material with aggregate in *asphalt*, like cement in concrete, and can be impregnated in a fibrous base to form *bituminous felt* for roofing and damp-proof courses. It can also be used as an emulsion in water for paints and protective coatings, in solvents for paints and adhesives, and as a hot-dip protective coating for metals.

A principal natural source of bitumen is *Trinidad lake asphalt*, which is a mixture containing about 40% of crude bitumen together with equal parts of sand and water. After digging, this is heated to about 150°C, to expel moisture and allow the coarser particles to settle. The refined material, known as Epuré, is stored in drums.

ASPHALT

Other, less rich, natural deposits of bitumen are found, mainly in Europe and Sicily, impregnated in rocks (chiefly limestones), and these are described as *natural rock asphalts*. After initial crushing, the rock is broken down to pass a $\frac{1}{8}$ in (3·18 mm) mesh sieve in rotating metal cages.

Rock asphalt is used mainly as a *road surfacing* material. It is first heated to 120–150°C to expel moisture, then reheated and spread hot to be tamped and rolled. A proportion of grit or sand may be incorporated in the rock asphalt.

Rock asphalt may also be used for *roofing* or *jointless flooring*.

Mastic asphalt

This is prepared by adding fine aggregate (mainly passing a BS 7/2·4 mm sieve) such as powdered limestone or rock asphalt, to an *asphaltic cement*, that is, heated bitumen (Epuré), lake asphalt, or a blend of these, and where necessary incorporating an oil as flux, with a proportion of grit as coarse aggregate (mainly less than $\frac{1}{8}$ in or 3 mm particle size). Pigments may be included for colouring.

The mixture is cast into blocks, commonly $\frac{1}{2}$ cwt (25 kg) and of cylindrical form, to facilitate storage and transporting.

For *roofing, jointless flooring, tanking* and *damp-proof courses* the cakes are broken up, heated to about 180–200°C, spread, and finished with a wood float. For roads, a proportion of coarse aggregate, usually up to 25% of crushed rock, gravel or blast-furnace slag up to 2 in size, may be added; in addition to this, a surfacing, such as $\frac{1}{2}$ or $\frac{3}{4}$ in (13 or 19 mm) granite chippings, may be applied and rolled in.

TAR MACADAM

Tar macadam is a road surfacing material consisting of coal-tar pitch and aggregate.

PITCH MASTIC

Pitch mastic is a jointless flooring or damp-proof coarse material consisting of aggregate with a binder of coal-tar or a coal-tar/bitumen mixture, laid hot at about 150°C. This flooring is liable to indent under heavy point loads.

There are a number of British Standards relating to asphalts for building purposes, and separate specifications apply to natural rock asphalts, as distinct from other types.

Requirements relate to the various constituents, such as the fine and coarse aggregate, flux oil and bitumen content.

BS 598 'Sampling and examination of bituminous mixtures for roads and buildings' includes methods of tests for asphalts, tar macadam and pitch mastic. It also covers determination of tar or bitumen content, water content, apparent specific gravity and voids, hardness, and examination of the soluble bitumen and aggregate contents.

BS 3235 'Test methods for bitumen' includes provision for the following tests, to be made under standard conditions.

1 Determination of ash after ignition
2 Ductility, as the distance by which a standard briquet can be elongated before breaking
3 Flash point (vapour ignition temperature) and fire point (temperature at which the bitumen burns)
4 Loss in weight due to heating (at 163°C for 5 hours)
5 Penetration (test for consistency) using a standard needle
6 Fluxing value of a flux oil (the effect of adding the oil on test results for Sections 4 and 5 above)
7 Softening point (ring and ball test)—the temperature at which a heated disc of the sample, held in a ring, allows penetration of a $\frac{3}{8}$ in (9·53 mm) diameter steel ball to a specified depth
8 Solubility in carbon disulphide
9 Viscosity of cutback bitumen and road oil, using a standard viscometer
10 Determination of water content

ROOFING FELTS

Roofing felts are manufactured in two grades, *saturated* felts and *coated* felts. Saturated felts consist of a fibrous base impregnated with a proofing material, which can be bitumen or pitch. They are not regarded as being impermeable. Coated felts are given the same processing initially as saturated felts, but have a subsequent application of a tough bitumen composition which makes them impermeable, and also allows them to be sanded or mineral-surfaced.

The fibrous base may be derived from rags, cotton, jute, flax, animal hair, wool, or asbestos or glass fibre. Glass fibre bases are normally resin-bonded, and are not therefore saturated but merely coated.

Felts are normally supplied in 12 yd/10 m and 24 yd/20 m rolls, 1 yd/1 m wide and specified by weight per roll. They are produced in four main types; self-finished, sanded, mineral-surfaced and reinforced.

Self-finished felts include saturated and coated grades, dusted with talc to prevent sticking when rolled. The saturated self-finished type is used as an underfelting for roofs. Coated types have a smoother surface finish, and, being thicker, are heavier. They are used as underfelting and in built-up roof coverings; two or three layers of felt laid breaking joint, sandwiched with liquid bitumen as adhesive, and the top surface sanded, or alternatively a mineral-surfaced felt may be used as the final layer.

Sanded felts are coated and sanded both sides to improve weathering and prevent sticking in the roll. They are used for roofing as a top layer or a lower layer.

Mineral-surfaced felts are coated felts finished on one side with a granular material such as crushed slate, and usually sanded on the underside. This gives improved appearance and weathering properties. They are used for single-layer roofing and as the final layer of built-up roofing.

Reinforced felts are coated felts which have a layer of jute hessian embedded in one side as reinforcement, and are also available with a layer of aluminium foil lining the reinforced side to give reflective insulation. They are used as underfelting beneath tiling, especially on unboarded roofs.

Deterioration of roofing felts

Roofing felts are liable to deteriorate under the action of heat and sunlight. Softening causes the impregnated substance to penetrate the base material thereby exposing the fibrous base to weathering action. To counteract this, self-finished types can be given a coat of bituminous paint every few years.

BS 747 'Specification for roofing felts'. Deals with requirements for roofing felts of the types described in this chapter.

BITUMINOUS D.P.C. MATERIALS

The quality of these is dealt with in BS 743 'Specification for materials for damp-proof courses', which includes bitumenised materials with bases of hessian, fibre felt, asbestos and lead, as well as mastic asphalt alone. The flexible type materials are similar to plain roofing felts.

BITUMEN EMULSIONS

These aqueous emulsions are used as paints in damp-proofing and in road surfacings. They are usually applied cold by brushing or spraying, and their use does not depend on dry weather or dry surfaces.

PITCH FIBRE PIPES

These are made by impregnating a fibrous base with pitch. The stages in manufacture include the mixing of the pulp stock (mainly cellulose fibre and water), and the removal of excess water by vacuum-isation then winding the stock on to a fine-mesh wire cylinder. The stock is then oven-dried and impregnated under pressure. The pipe is cooled in water and its ends are finally tapered.

Couplings are cut from larger diameter pipes and internally tapered. Other fitments may be in moulded plastics. The pipes are dry-jointed, as they are simply driven together, and may be cut using a handsaw.

These pipes are used for soil and surface water, and are also available for above-ground work (e.g. soil stacks and connections to fitments). In addition, there are perforated pipes for land drainage.

Pitch fibre drain and sewer pipes are dealt with in BS 2760.

MAGNESIUM OXYCHLORIDE (MAGNESITE) FLOORING

When an aqueous solution of *magnesium chloride* is mixed with powdered *magnesium oxide*, a chemical reaction takes place which results in solidification. This is the principle on which magnesium oxychloride jointless flooring is based. Commercial grades of the two materials are used, together with added fillers and possibly a pigment.

The magnesium chloride solution is prepared by covering the solid with water and stirring it periodically until it dissolves. The solution obtained is diluted with water to achieve a specific gravity of about 1·17 for a single coat or top coat, or 1·15 for a bottom coat. One- or two-coat treatment can be used.

The magnesium oxide powder is mixed with between one-half and equal parts by weight of filler, depending on the type of work and whether it is a bottom or top coat. For a bottom coat ($\frac{1}{2}$–$\frac{3}{4}$ in/13–25 mm thick) equal parts of magnesium oxide and sawdust are normally used. For a top coat ($\frac{1}{4}$–$\frac{1}{2}$ in/6–13 mm) or a single coat ($\frac{1}{2}$–$\frac{3}{4}$ in/ 13–19 mm) the smaller amount of filler would be needed, consisting usually of a blend of wood-flour and pigment, together with powdered silica if greater wear-resistance is needed.

The mix is prepared by adding just enough magnesium chloride solution to produce a consistency to allow for tamping in the case of a bottom coat, or slightly more for a top coat to give a stiff consistency suitable for spreading by trowel and tamping.

An excess of calcium chloride solution, or an overstrength solution, is likely to give rise to persistent dampness in the floor since it is hygroscopic. It is also corrosive to metals, so all metal must be protected from attack by a bitumen or other suitable coating.

FLEXIMER FLOORS

These are various types of flooring consisting of a rubbery, resinous or bituminous binder together with aggregate, pigment and fillers. They include emulsion type binders, such as natural and synthetic rubber latex, aqueous bitumen emulsion and aqueous PVA emulsion. Others are based on synthetic resins, which may be hardened by adding a catalyst.

LATEX-CEMENT FLOORING

This is a fleximer floor based on ordinary Portland or high alumina cement, aggregate, fillers and pigment, gauged with aqueous emulsion of rubber latex. The rubber latex forms between 10% to 25% of the total mix, by weight (increasing for greater wear-resistance and reduced permeability).

Aggregate may be crushed stone, sand, granulated cork, vulcanised rubber or hardwood chips. Fillers include wood flour, powdered asbestos, slate, limestone or French chalk.

The usual thickness of the finish is $\frac{1}{4}$–$\frac{1}{2}$ in (6–13 mm).

Summary

Bitumen and asphalt

Asphaltic bitumen is a naturally occurring hydrocarbon, or petroleum derivative, solid at normal temperatures, but thermoplastic. It is used as a binder in asphalt, as a hot-dip coating for roofing and d.p.c. felts and pipes, and in bitumen paints and emulsions.

Principal sources of bitumen are Trinidad lake asphalt, and rock asphalt deposits (limestone impregnated with bitumen). Rock asphalt, used for road surfacing, roofing and jointless flooring, is made by crushing the impregnated rock to pass a $\frac{1}{8}$ in (3·18 mm) mesh sieve, and heating it, together with a proportion of grit or sand if required.

Mastic asphalt is a blend of asphaltic cement (Epuré) and powdered limestone or rock asphalt (together with a fluxing oil, grit and pigment where required). It is also used for roofing, jointless flooring, tanking, as a d.p.c. material and for roads, often with added coarse aggregate.

Tar macadam is a road surfacing material of coal-tar pitch and aggregate.

Pitch mastic, used as jointless flooring and as a d.p.c. material, consists of an aggregate with coal-tar or coal-tar/bitumen binder.

Roofing felts may be saturated or coated. Varieties include self-finished, sanded, mineral-surfaced and reinforced felts.

Bitumen felt is used as a flexible d.p.c., and mastic asphalt is used as a semi-rigid d.p.c.

Bitumen emulsions are used in paints, in damp-proofing and in road surfacing.

Pitch fibre pipes are used for soil and surface water, and for above-ground work.

Magnesium oxychloride (magnesite) flooring is made by mixing an aqueous solution of magnesium chloride with a dry mixture of magnesium oxide, fillers and pigment.

Latex-cement flooring is a fleximer floor based on ordinary Portland or high alumina cement gauged with rubber-latex.

17

GLASS

Ordinary glass is a soda-lime-magnesia silicate, prepared by heating a mixture of sand (silica), soda (sodium carbonate), limestone (calcium carbonate) and dolomite (magnesium calcium carbonate). Together with certain other ingredients and some broken glass, called 'cullet', the batch ingredients are melted in a furnace.

If sand and soda are mixed and heated, the soda acts as a flux and the sand dissolves in it. With excess of soda the resulting liquid remains syrupy on cooling, but with more sand a hard, transparent substance is obtained. This substance, known as *water-glass*, is soluble in water, but by including in the original mixture a suitable amount of lime and dolomite a glass which is not water-soluble results.

The method of manufacture is such that cooling to the hardened state is fairly rapid, during which the glass passes through a viscous stage when it can be worked to the required shape. This rapid cooling is necessary to prevent the glass from undergoing crystallisation, or *devitrification*, with loss of transparency and strength. The final cooling stage, after hardening, must be gradual to prevent cracking, and is referred to as *annealing*.

The glass may at any time be re-softened by heating, to allow further working, if required, provided this is done in such a way that devitrification is prevented.

Variation in the detailed procedure of manufacture, and the inclusion of special ingredients, results in different types of glass.

SPUN GLASS

Spinning is an older process which has survived mainly for the production of the traditional 'bull's eye' window panes—originally known as 'crown' glass. A bulb of molten glass is formed at the end of a pipe by blowing into it. The lower end of the bulb is gathered on to an iron rod, the blow-pipe is cut off, and the remainder is rotated (spun) to the form of a disc.

CYLINDER BLOWN SHEET GLASS

This process is no longer in general use but is of historical interest.

A blown bulb of glass was extended, cut at its lower end and rotated to cylindrical form. When cooled it was cut lengthwise, reheated and flattened to sheet form, then annealed.

CYLINDER DRAWN SHEET GLASS

In this method, which has been superseded by the flat-drawn processes, molten glass was drawn vertically from a pot by the lipped end of a pipe to form a cylinder some 30 in in diameter and 40 ft high. When cooled, the cylinder was cut lengthwise into several sections, each of which was flattened after reheating, then annealed by being passed through a lehr (a long tunnel, hot at one end and cool at the other).

FLAT DRAWN SHEET GLASS

In a more recent, *flat drawn*, process the molten glass is first heated in a large tank to over 1500°C to achieve the correct chemical formulation, then cooled to 1000°C, when it is sufficiently viscous to be drawn vertically to form a continuous sheet passing between asbestos-covered rollers. The leading edge of the sheet is initially attached to an iron grille or 'bait' which is dipped into the molten glass, then slowly raised. The sheet rises to a height of 30 ft (9 m), passing through a tower of water-cooled chambers in which it is annealed. On emerging at the top it is cut off to the required length, and trimmed.

The surfaces of the glass are never perfectly flat, resulting in slight distortion of vision, but have characteristic brilliance known as a 'fire-finish'.

ROLLED GLASS

The rolling method is used principally in the production of figured glasses—that is where one or both surfaces are textured or impressed with a pattern, which prevents clear vision through the glass.

In the original process the molten glass was poured on to a metal table in front of a roller which traversed it at a height to produce a sheet of the required thickness. Any pattern on the roller is impressed in the surface of the glass.

The more recent system is to draw a sheet horizontally between double rollers.

Types of glass which can be made by this method include those known as *rolled, figured rolled, rough cast, cathedral* and *wired*.

POLISHED PLATE GLASS

The demand for a glass with perfectly flat, parallel surfaces, allowing undistorted vision, led to the process of grinding and polishing rolled sheet glass. At first only one surface was ground at a time, but a later development was to pass a continuous ribbon of glass straight from rollers to a machine which ground both surfaces simultaneously.

The expense of grinding and polishing led to attempts to produce glass with equally true surfaces by other methods, and this resulted in the 'float glass' method.

FLOAT GLASS

In this method, molten glass is floated on to the surface of molten tin, then allowed to cool. This produces glass having flat, parallel surfaces and the fire-finish associated with drawn sheet glass, so the method has replaced the polished plate method. However its present costs of operation are such that it has not yet replaced the drawn sheet process.

PRESSED GLASS

In this method the molten glass is pressed into a steel mould by a plunger. It cools immediately, is removed and annealed. In this process the composition of the glass is varied to increase the working temperature range to compensate for the more rapid cooling. Hollow glass blocks (glass bricks) are a typical example of a product made in this way (formed in two halves then sealed together to give a hollow core). Their patterned surface makes them non-transparent, but they are translucent and so can provide illumination by transmitted light.

COLOURED GLASS

Colour may be 'flashed' in a thin, integrally formed surface layer on to the clear sheet during its manufacture. Alternatively it may be introduced in the body of the molten glass to form a uniformly tinted sheet, when it is known as 'pot' colour.

OPAL GLASS

Opal glass contains countless minute crystals (often sodium and calcium fluorides), formed usually by adding special ingredients during manufacture and varying the rate of cooling of the glass. Opacity results from the difference of their refractive index from that of the glass matrix, and varies in degree according to their concentration. Natural opalescence is white, but opal glasses may also be coloured.

SPECIAL PURPOSE GLASSES

There are many types of glass produced with special properties to meet particular needs, some of which must be mentioned.

Toughened glass

Ordinary glass is strong in compression but weak in tension and consequently has the disadvantage of being fragile. Deflection of a sheet of glass due to a bending or impact load results in one face of the sheet being put in compression and the other face in tension, and breakage will occur if the comparatively low tensile strength of the glass is exceeded. However, the glass may be considerably strengthened by subjecting it to a special heat treatment. This tempering process consists of chilling the surfaces of the glass during the cooling stage of its manufacture, or after re-heating, by blasts of cold air. The glass contracts on cooling, and the differential rate of cooling throughout its thickness leaves the sheet with its two surfaces in compression and the interior in tension (a form of pre-stressing). The effect is to greatly increase the load which the glass can sustain.

The toughening process is also applied to some pressed or moulded glass units like those which are used in pavement lights and roof light construction.

If the surface of toughened glass is damaged the internal stresses are suddenly unbalanced and the whole of the sheet disintegrates—but into small non-splintering fragments which do not constitute the danger attributable to ordinary broken glass.

Heat resisting glass

This may be either toughened glass, which has also considerable resistance to thermal shock, or glass having a specially low thermal expansion coefficient.

Heat absorbing glass

This is glass which is substantially opaque to solar 'radiant heat' (infra-red) but which transmits most of the visible light. This glass usually has a bluish-green tint.

Wired glass (Fig 71)

This is glass containing wire mesh embedded centrally in the sheet during manufacture which holds it together when broken. This makes it useful in meeting requirements for safety—especially in roof-lights—and in resisting the spread of fire. Common types are *Georgian* ($\frac{1}{2}$ in/13 mm square mesh, electrically welded), *Diamond* ($\frac{3}{4}$ in/19 mm sided, electrically welded) and *Hexagonal* ($\frac{7}{8}$ in/22 mm size mesh, entwined at intersections). Wired glasses are obtainable either transparent (polished) or translucent (cast).

Lead (X-ray) glass

This glass is specially formulated to have a high level of absorption to X-rays.

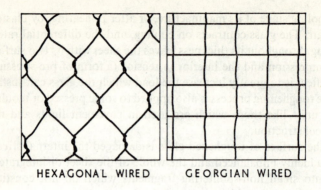

HEXAGONAL WIRED GEORGIAN WIRED

DIAMOND WIRED

Note: these diagrams are at reduced scale

Fig 71 Wired glass

Safety glass

This may be either toughened glass, or glass of laminated construction with a special core of plastics.

CLASSIFICATION AND GENERAL PROPERTIES OF GLASS

Thickness

The thickness of drawn sheet glass is denoted by its mass per unit area—called its *substance*. The common substances for flat drawn sheet glass for windows, together with their *approximate* equivalent thicknesses are 24 oz/7·32 kg/m² ($\frac{1}{10}$ in/2·5 mm), 26 oz/7·93 kg/m² ($\frac{1}{8}$ in/3 mm) and 32 oz/9·76 kg/m² ($\frac{5}{32}$ in/4 mm. Imperial substances stated in ounces relate to 1 ft² area.

Other types of glass are normally specified by nominal thickness. Thick drawn sheet glass is obtainable in thicknesses $\frac{3}{16}$ to $\frac{1}{4}$ in (5–6·5 mm). Float glass is currently available in thicknesses of $\frac{1}{8}$, $\frac{3}{16}$, $\frac{1}{4}$ and $\frac{3}{8}$ in (3, 5, 6, and 10 mm). The thickness for rolled glasses may be $\frac{1}{8}$, $\frac{3}{16}$, $\frac{1}{4}$ or $\frac{3}{8}$ in (3, 5, 6 or 10 mm) according to type, with the standard for wired glass $\frac{1}{4}$ in (6 mm). Plate glass is obtainable in various thicknesses up to 1 in (25 mm) and above. All metric equivalents stated are approximate.

Classification

BS 952 'Classification of glass for glazing and terminology for work on glass' provides a useful reference to the various types of glass for glazing within the following categories: transparent and translucent glasses for general purposes, opal glasses and glasses for special purposes.

The BS designates various qualities of sheet and plate glass by a letter-code system which is in common use in the trade—see Table 36, page 324.

Staining

A dull white stain will often form on glass sheets left for a considerable time with their surfaces in contact when wet. This is due to the leaching of sodium ions from the glass, which form caustic soda with the trapped water, giving a resultant solution which can attack glass surfaces.

GLASS WOOL

Fibrous forms of glass, used mainly as a reinforcing material, or for heat insulation and fireproofing, are made by drawing, or blowing, molten glass from a perforated plate. The thickness and length of the fibres may be controlled in manufacture to make them suitable for different purposes, and they may be spun together. The names *continuous filament glass fibre* and *glass wool* insulation are commonly used to describe these.

Summary

Glass is a soda-lime-magnesia silicate, cooled rapidly from the molten state in manufacture to prevent its crystallisation, and so obtain transparency.

Before cooling to the hardened state it passes through a plastic condition, during which it can be worked. It can be re-softened by heating.

Processes for the manufacture of window glass have included the spun glass, cylinder blown sheet, cylinder drawn sheet and flat drawn sheet methods. The flat drawn sheet method gives a 'fire-finished' surface, with slight distortion of vision.

A different method is used for rolled glass (e.g. rolled, figured rolled, rough cast, cathedral and wired).

Polished plate glass, giving undistorted vision, is made by grinding and polishing rolled sheet.

Float glass, made by floating molten glass on molten metal and allowing them to cool, can produce glass which combines the surface brilliance characteristic of fire-finished glass with the undistorted vision of plate glass.

Pressed glass products include translucent hollow glass blocks (glass bricks).

Coloured glass is made either by 'flashing' a thin layer of coloured glass on to the clear sheet during manufacture, or by colouring the whole body of the molten glass ('pot' colour).

Opal glasses are rendered non-transparent by their content of minute crystals, caused by the addition of special ingredients and varying the cooling rate during manufacture.

Special purpose glasses include toughened (heat treated) glass, heat resisting glass, heat absorbing glass, wired glass, lead (X-ray) glass, and safety glass.

Drawn sheet glass is specified by its substance (mass per unit area), and other types normally by their nominal thickness.

Staining due to leaching of sodium ions to form caustic soda can result from wet sheets of glass being kept in contact.

Glass fibre for insulating and reinforcing purposes is made by drawing or blowing molten glass from a perforated plate.

18

METALS AND THEIR PRODUCTS

EXTRACTION OF METALS

Most metals are found in the Earth's crust combined with other substances as compounds known as *minerals*. The most common of these mineral compounds are the oxides, carbonates and sulphides of the metallic element. The minerals are themselves usually mixed with earthy impurities, or rock, forming together what is known as an *ore*.

When metal is to be extracted from the ore, the mineral must be first separated from the earthy impurities; for example, by crushing it and washing out the less heavy impurities. This is followed usually by some form of heat treatment to break down the mineral into its constituent elements, although other methods may be applied in particular cases. Examples of these are given later in this chapter. When roasted, the volatile impurities (e.g. sulphur, phosphorus and arsenic) are driven off and the metal itself is usually converted to an oxide. The oxygen can be removed by a *reducing agent*, usually heated coke (carbon), which itself combines with the oxygen. The remaining earthy impurities usually combine with an added *flux*, often limestone, to form a fluid crust at the surface of the molten mass, known as *slag*, which can be run off. In some cases the impurities of the ore itself act as the fluxing agent. The main purpose of the flux is to lower the melting point. After being extracted in this way the metal will still contain small amounts of impurities, and may be reheated to remove these, that is, *refined*. However, certain minor impurities may impart valuable properties to the metal, so they are not removed in all cases. In fact, small amounts of particular 'impurities' may be added on purpose to give rise to an *alloy* with some desired special property such as increased strength, hardness or corrosion resistance. The entire process of extraction is termed *smelting*.

A rich ore is an ore which has a high content of metal, and the cost of extracting the metal from an ore depends partly on its richness, but also on the complexity of the process necessary. For example, the smelting of iron is relatively cheap due to the simplicity of the process, and because the ore used contains upwards of 70% of the metal.

Iron and its alloys, including steel, are known as *ferrous* metals.

THE WORKING OF METALS

The process of bringing the extracted metal to the required finished shape of the product is known as *working* the metal. Products may be either cast or wrought.

A *cast* product is a product which is obtained by pouring molten metal into a mould of the required shape and allowing it to cool. Subsequent processing of the solidified mass (the *casting*) is limited to surface finishing operations such as machining, or drilling.

A *wrought* product is a product which is obtained by forcibly shaping solid metal to the required form by various alternative methods such as *rolling* (e.g. to bars, sheet, strip or other simple sections), *forging* (hammering), *pressing* (from sheet), *drawing* (e.g. into wires or tubes), or *extruding*, that is, forcing a hot mass of metal through an orifice which has the shape of the required section; for example, structural sections, rods, strips, pipes, tubes, gutters, door and window frame sections and glazing bars.

If the simplicity of the basic manufacturing process had to be proved a cast article would be preferable to a wrought article; in practice, however, the casting has its limitations with regard to size and complexity of shape; in addition to this, the working required to obtain a wrought part can also improve certain mechanical properties and sometimes eliminate defects such as pinholes.

The processes of welding and soldering are useful as ancillary methods of fabrication. Welding consists of joining metal parts by their fusion. In *soldering*, a metal or alloy (the solder) of lower melting point than the metals to be jointed is melted in contact with their butted or lapped surfaces and allowed to solidify. The common *soft solders* are lead-tin alloys. *Brazing* is a form of soldering in which the jointing alloy is a brass or other high-melting point alloy known as a *hard solder*.

Of course a particular metal or alloy may be more suited to one method of working than another, with regard to its mechanical properties such as *malleability* (a malleable metal is one easily hammered into thin sheets); *ductility* (a ductile metal is easily drawn into wires); *toughness* (resistance to fracture by dead, live and impact loads), and *hardness* (resistance to surface indentation or scratching).

TESTING METALS

In the research and development fields, the range of testing which is carried out is unlimited and results in frequent improvement of metals already available, and in the production of new special types to meet specific user demands.

In contrast, the requirement in the *commercial testing* of metals is to standardise, and simplify where possible, tests used to check specific basic properties in order to maintain quality control in production, or to check the metal's compliance with a particular specification to establish its suitability for a given purpose, or to check for deterioration in a used part.

The mechanical properties of a metal may be altered during preparation of a test specimen, for example, by work-hardening. Therefore, special conditioning, usually by heat treatment, may be required before testing to bring the metal of the test specimen to its original, or some standard, condition.

An outline of some standard tests follows.

Tensile test

This test is usually applied either to a cylindrical test piece or flat bar or strip cut from the metal in question. The specimen is stretched by loading it progressively in a testing machine until fracture occurs.

In addition to measuring the load at fracture, which allows calculation of the ultimate stress, the amount by which the specimen elongates over the marked *gauge length* before fracture may be determined. This is a measure of the metal's ductility. The reduction in area at the point of fracture (see Fig 72) is another measure. A *brittle* (non-ductile) metal fails suddenly, with very little warning by elongation.

Results for the tensile test are calculated as follows:

$$\text{yield stress} = \frac{\text{yield load}}{\text{original area of section}}$$

$$\text{ultimate stress} = \frac{\text{ultimate load}}{\text{original area of section}}$$

$$\text{elongation, per cent} = \frac{\text{elongation}}{\text{original gauge length}} \times 100$$

$$\text{reduction in area, per cent} = \frac{\text{contraction in area of section}}{\text{original area of section}} \times 100$$

$$\text{Young's modulus (E)} = \frac{\text{stress}}{\text{strain}} \quad \begin{array}{l}\text{(within limit of proportionality, see} \\ \text{Volume 2)}\end{array}$$

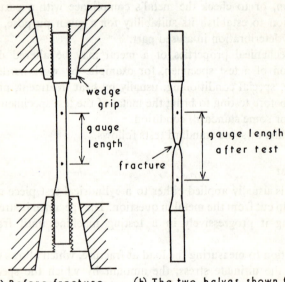

(a) Before fracture   (b) The two halves shown fitted together after fracture

Fig 72 Tensile test piece for steel

Cold bend test

This is a simple test which gives an indication of a metal's ductility. It is also a practical guide to a metal's suitability for wrought work. The test, used mainly for metals in bar or strip form, consists of doubling the piece over on itself to a 'U' shape, in the unheated condition. Specifications usually require the test piece to show no surface fracture after forming to a given internal radius of bend.

Hardness test

Tests commonly used are the *Brinell*, *Rockwell* and *Vickers* methods, which measure the resistance offered by the metal to the penetration of a hardened steel ball or a diamond into its surface under standardised loading conditions.

For the Brinell test a result is expressed as a number, obtained by dividing the load applied (kgf) by the area of the impression (mm^2) caused by a 2 mm diameter ball. The test is suitable for metals of hardness up to 500. For harder metals a diamond indentor is required. The Rockwell test is similar to the Brinell method but measures the depth of impression of a steel ball or diamond. The Vickers method is again similar but a diamond indentor is used and the depth of impression is measured.

Impact test

In the *Izod* and *Charpy* methods of standard test a notched specimen is struck by a falling pendulum and the energy expended in breaking the specimen is indicated by the height to which the pendulum rises at the other side of its swing.

Microstructure

Reference to the *grain structure* of materials was made in Chapter 6 of Volume I, where it was seen that this relates largely to such factors as the rate of cooling of the melt and any heat treatment to which the solid is subjected. With metals it depends also on any impurities present.

Whereas the grain structure of many igneous rocks is easily seen without magnification, this is not the case with metals. An exception occurs in the special case of the zinc coating of some hot-dip galvanised products which may clearly show large grains of zinc. In

K

most other cases, the grain structure of metals can only be studied with the aid of a microscope, and using a specimen which has been specially *etched* to cause the grains to show up under reflected light. Etching is done by applying a chemical solvent to the surface of the metal. The solvent action is more effective at the grain boundaries than elsewhere. In addition to this, the surface of the individual grains is left with different reflection characteristics, so that they are distinguishable. Under the microscope they will then have characteristics of the type seen in the case of the zinc coating, which tell the metallurgist something of the nature, history and properties of the metal. For example, finer grains usually indicate greater strength and hardness. The etching solution varies with the metal. For steel a mixture of 98 parts of alcohol and 2 parts of nitric acid can be used.

In the case of *alloys* more complex microstructures can occur, according to which metals are present and in what proportions. In some cases, grains of the individual constituents may be distinguished. This commonly occurs in the iron-carbon alloys (steels). In other cases, for example, the copper-nickel alloys, the grains may appear unlike those of the constituents. A new grain structure, characteristic of a single metal, is formed, in which case the alloy is described as a *solid solution*. Many other binary alloys form a solid solution only at one particular composition, and at all other compositions have a microstructure of two different grain forms. In yet another category are the alloys which form eutectics, such as the lead-tin alloys (solder). The eutectic composition for this alloy (see Volume 1) is approximately 37% lead and 63% tin. If more tin is present, the microstructure will be of tin and eutectic. If there is more lead, a structure of lead and eutectic will occur.

IRON AND STEEL

Smelting of iron

The main ores from which iron is extracted are those ores which contain the oxide or the carbonate of the metal. Extraction is usually in a *blast furnace* of steel construction lined with firebrick, about 100 ft (30·48 m) high and 25 ft (7·62 m) wide. This is charged continuously with layers of ore, coke and limestone or other flux if needed.

The fuel is kept burning, and hot air is forced in at the bottom, where the maximum temperature (1790°C) is reached.

The volatile impurities are driven off, and carbon and carbon monoxide reduce the iron oxides. Non-volatile impurities (e.g. silica and clay) combine with the flux to form *blast-furnace slag*, which floats on the molten metal and is run off at intervals. Impure iron is drawn off at the base and cast in rectangular blocks ('pigs'). This pig iron contains impurities, including up to about 5% of carbon, which may be present either combined with the iron as *ferric carbide* in a compound form known as *cementite*, or uncombined in the soft graphite form of the element, depending on the rate of cooling the molten metal. If impurities are present mainly as cementite (rapid-cooled) a brittle *white pig iron* results, whereas if they are present mainly as graphite (slow-cooled) a non-brittle *grey pig iron* is formed. *Mottled pig iron* contains both combined and uncombined carbon. These three names refer to the appearance of the metal on fracture.

Pig iron is used in the production of cast iron, wrought iron and steel. These processes will now be considered.

Iron castings

Cast iron products, as the name suggests, are cast direct from molten metal, either in *sand moulds* or in *dies* (metal moulds, which give rise to the term *die-casting*). A special method which can be used for hollow castings, such as pipes, is the centrifugal or *spinning* process.

Cast iron

This is made by remelting grey pig iron together with a proportion of scrap in a *cupola*, which operates like a small blast furnace. Coke is used as fuel and limestone is used as a flux. By this process certain impurities are removed as slag, and free carbon is combined. The extent to which this combination occurs largely determines whether a very hard and brittle *white cast iron* (low free carbon) is obtained with low tensile strength, or a somewhat less hard and brittle *grey cast iron* (high free carbon) of higher tensile strength. Both forms have extremely high compressive strength and wear resistance, together with remarkably high resistance to rusting.

Cast iron is widely used, mainly in the grey form, since this expands slightly on solidifying and is therefore suited to casting, whereas the

K*

white form contracts. It is used for cast rainwater and soil goods (e.g. pipes, gulleys and manhole covers), pipes for gas and water, baths, boilers, stoves, flue pipes and gratings.

Malleable cast iron

This type of cast iron, markedly more ductile and shock resistant than ordinary cast iron, is made from white cast iron by a heat treatment (*annealing*) which modifies the form of the carbon content. The resulting grey cast iron is used for baths, hinges and for malleable cast iron pipe fittings for steam, air, water, gas and oil.

Special cast iron

Special grades of cast iron are produced by alloying with, for example, chromium, to give increased hardness, or with copper to improve corrosion resistance.

Wrought iron

This is produced by refining pig iron so that it contains less than 0·15% of carbon. The pig iron is remelted by heating it to 1200°C in a small *reverberatory* furnace, in which the heat is applied by passing the hot gases from the fire over the surface of the charged hearth. This results in the oxidation of carbon and other impurities to form a slag. The melting point of the purified iron is higher than the melting point of the impure form, and the furnace temperature, so it solidifies and is extracted as white-hot lumps. The resulting mass of mixed slag and iron is hammered and rolled to expel the slag, and this action improves the strength and ductility of the metal.

Wrought iron is easily worked and welded because of its ductility, but its use is now limited due to the improved and cheaper methods available for the working of mild steel which is an alternative material. Its natural resistance to corrosion makes it superior to steel in this respect, but effective protection of steel against corrosion is easily secured, for example, by coating or wrapping methods. It has been widely used for decorative wrought iron work, pipe fittings, hinges and bolts.

Steel

Steel is an alloy of iron, containing elements such as carbon and manganese. Its carbon content is intermediate between those of cast and wrought iron, but is varied within this range according to the grade of steel required. For example, a rough differentiation is made between *low carbon* steels (0·1 to 0·3% approximately) such as ordinary *mild* steel, and *high carbon* steels (1 to 2% approximately) such as *tool steels*. Both are known as *plain carbon* steels, to distinguish them from steels containing alloying ingredients additional to carbon which gives them special properties.

The method of manufacture at present widely used in Great Britain is the Siemens-Martin *open hearth process*, in which pig iron, together with controlled proportions of scrap and ore, are fed into the hearth of a reverberatory furnace and heated to a very high temperature by firing carbon monoxide gas, which is blown across the surface together with blasts of hot air. The oxidising flame removes impurities. The metal is run off and moulded to ingots. It may later be hot-rolled, pressed or hammered to the required shape, that is, *forged*. It is a special characteristic of mild steel that it is sufficiently ductile to be readily forged, whereas high carbon steels are normally too hard for this. Less frequently, the metal is cast direct to the finished shape as *cast* steel, which provides a suitable alternative to cast iron in cases where greater tensile strength, or less brittleness, is preferred. On the other hand, the process of forging steel improves its tensile strength and ductility.

Other methods of steel production include the *Bessemer process*, in which the impurities are removed from the molten pig iron by blowing air through it to oxidise (burn) them; an advantage is that no fuel is needed.

Steel of high purity, or high-grade alloy steels, can be obtained using the electric arc furnace, in which an electric arc passes from carbon electrodes causing the metal to melt, or to be further heated if already molten when introduced. An important feature is the ability of this process to operate entirely on scrap metal.

Steel alloys

These are produced by adding one or more elements to steel. By this means we obtain steels with special properties such as structural *high-tensile* steel, the intensely hard, yet tough, *manganese* steel used for rock-crushing and machinery components, *tungsten* steels for tools with high-speed cutting edges, the non-rust *stainless* steel (a chromium-steel alloy) and the resilient *spring* steel (a silicon-steel alloy).

Steel in building

Steel is used for innumerable purposes in building, often plated with other metals (e.g. chromium, nickel or tin), galvanised (zinc coated) or enamelled. Products include structural steel sections; reinforcing bars, wire or mesh for reinforced concrete; metal lathing; pipes and fittings for soil, waste, water, gas and steam; electrical conduit, rainwater goods, boilers, cylinders, tanks and cisterns; lavatory basins, sinks and baths; corrugated sheets (galvanised); doors, windows and metal trim (e.g. skirtings and rails); nails, bolts and fittings. Stainless steel is used for some lavatory basins, sinks, drainers, boilers, baths and bath panels and curtain walling. Chromed fittings are popular for plumbing fittings and accessories.

NON-FERROUS METALS

Aluminium

Among the metallic elements, aluminium occurs in greatest quantity in the Earth's crust, mainly in the form of hydrated aluminium silicates and oxides. These compounds are the main constituents of clays, formed by the weathering of igneous rocks. Unfortunately the extraction of aluminium is only readily achieved from the hydrated oxides, whereas ordinary clays contain mainly the hydrated silicates. However, clays rich in aluminium oxides do occur, and they are given the name *bauxite*.

It is not possible to obtain aluminium in a purified form using the traditional smelting methods, and its extraction only became a commercial proposition when an *electrolytic method* was introduced in 1886. This has two main stages. The first is the heating, under pressure, of crushed, powdered ore in strong *caustic soda* solution, in which the aluminium oxide dissolves but the impurities do not. The impurities are filtered off from the solution; the remaining *aluminium hydroxide* is obtained by precipitation and further filtration. Heating removes the combined water, leaving pure *alumina* (aluminium oxide). The second stage is the splitting up of alumina into its elements by passing a direct electric current through it (Fig 73). The alumina, together with cryolite (sodium aluminium fluoride) as flux, is placed in an iron bath lined with carbon forming the *cathode*. Carbon electrodes which dip into the bath form the *anode* of the cell, and when a current is passed the temperature of the charge rises to about 1000°C causing it to melt, and it releases its oxygen at the anode. The metal which is liberated at the cathode collects at the bottom and is run off for casting to ingots.

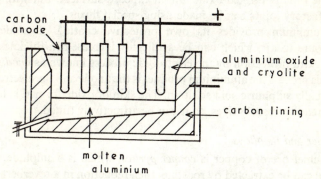

Fig 73 Extracting aluminium by the electrolytic method

Properties and uses of aluminium

The important properties of aluminium include its light weight (about a third of that of steel), good conductivity to heat and electricity and high resistance to corrosion when free of impurities. However, it can be attacked by alkalis when damp, and is liable to suffer electrolytic corrosion when in contact with other metals, particularly copper. Its high reflective properties make it useful as a barrier to radiant heat transmission (e.g. in *foil* form).

The pure cast metal is comparatively soft, and only of moderate tensile strength, but alloying with small amounts of other metals can considerably increase its strength and hardness to equal or exceed that of mild steel. Such alloys may be used for structural members, scaffolding and ladders. The metal's properties can also be modified by heat treatment or by working. Its ductility or softness enables the metal to be easily *drawn* into wires or *extruded* to various sections, such as structural members, pipes, tubes, gutters, door and window frames. Its malleability enables it to be rolled to thin sheets (aluminium foil). Hot-rolling produces sheets for roofing, cladding, etc., or strip used to form gutters, flashings, etc. The metal is also readily cast to shape.

Aluminium is produced in different grades of *purity*, e.g. sheets 99%, 99·5%, 99·8% pure, and super purity (99·99%), and it should be noted that the purer forms are the easier to work. *Alloying* gives increased strength but with reduced workability. Aluminium and its alloys are not easily welded, due to interference by the rapidly formed oxide film at exposed surfaces. But with care satisfactory joints can be made using special fluxes.

Aluminium provides its own protective coating of oxide on exposure to air, which can be artificially increased in thickness to give extra protection by the electrolytic process known as *anodising*. In this process the aluminium product is the anode and the electrolyte is usually sulphuric acid solution. The skin of oxide formed is hard, and readily absorbs dyes, so it can be attractively tinted.

Copper and its alloys

The chief ore of copper is *copper pyrites*, which is a sulphide. The metal can be extracted by roasting, then reduction in a reverberatory furnace. The resulting crude copper is refined by re-smelting, and for super-purity copper (required for electrical conductors) is further refined electrolytically.

Copper itself is a very useful metal and gives rise to a number of important alloys. The best known are *brass* (copper-zinc) and *bronze* (copper-tin.) Both are used for ornamental work, door and window furniture, and also very extensively as plumbers' brass-foundry and sanitary engineering fittings such as taps, valves, waste traps and pipe couplers (unions). Like copper, these alloys may be plated with chromium or nickel to give a very durable and attractive finish.

Copper is malleable, ductile and of fairly high mechanical strength. However, it hardens when cold-worked (e.g. by hammering, drawing or cold-rolling) and may have to be annealed by heating to red heat after such treatment to enable it to be bent or shaped. It has extremely high resistance to corrosion including alkalis but should be protected by bitumen-coating or binding as it can be attacked by acidic soils. The natural *patina* (skin) acquired on exposure to the atmosphere is attractive as well as protective, so that painting is unnecessary. It has high conductivity for heat and electricity and is easily welded, soldered or brazed.

Copper in building

The main uses of copper in building are for pipes and fittings, rain-water goods, boilers, tanks and cylinders, sheet and strip for roof covering and d.p.c., and electricity cables. It is also extruded as glazing bars and window frames.

Copper tube (pipes), sheet and rod are in general supplied in three tempers, *soft* (annealed), *half-hard* and *hard*. This should be given consideration when ordering to ensure the material's suitability for the purpose concerned. For example, soft temper is appropriate for roofing work for its ease of shaping, and is used for copper tubing supplied in coils for laying underground. Straight lengths of tube are used both above and below ground. Copper tube is supplied in light and heavy gauges; heavy gauges are used underground.

Lead

Lead is extracted from the mineral *galena* (lead sulphide) by roasting, followed by reduction in a reverberatory furnace. The resulting *pig lead* may be further refined by re-smelting or by an electrolytic method.

Lead is the most ductile and malleable of the common metals, and sheets are easily cut with a knife or shears, worked and jointed by such methods as soldering and welding. It is highly resistant to corrosion generally. It is very heavy (s.g. 11·3), has low tensile strength, high thermal expansion and a pronounced tendency to creep. Its strength can be increased, and creep can be reduced, by alloying, as in *tellurium lead* and *silver-copper-lead alloy*.

Uses of lead

It is widely used in sheet and strip form for roofing and damp-proof courses. Lead and its alloys are also used for pipes for water, gas and sanitation and for plumber's fittings such as waste traps.

Lead sheet and strip is produced mainly by rolling (milled lead), although thick sheets can be cast in sand moulds. These forms are used mainly as roof covering and damp-proof course material.

Lead pipes are formed by extrusion. Pipes up to 2 in (or 50 mm) diameter are usually supplied in coils, and larger sizes in straight lengths.

Zinc

Zinc is extracted mainly from the mineral *sphalerite* (zinc sulphide) known as *blende*. The traditional smelting process is rather unusual, because of the very low boiling point of zinc (907°C) in comparison to the other common metals and in relation to its own melting point (420°C).

After roasting the ore, then reducing the zinc oxide which has been produced, by heating it with coal dust, the vaporised metal is condensed and run into moulds. This commercial zinc or *spelter* is refined by re-distillation. Alternatively, an electrolytic method can be used to provide zinc of super purity.

Zinc is easily soldered, does not creep and is not prone to work-hardening. In pure form, or as alloys, it is highly resistant to corrosion, even, it is claimed, in marine atmospheres, due to the protective skin which forms on its surface through exposure to the atmosphere. Precautions should be taken to avoid electrolytic corrosion between zinc and other metals, especially copper.

Zinc in building

Zinc is widely used in sheet and strip form for such purposes as roofing and panelling. It is also used for wrought rainwater goods and light alloy fittings. Other uses are as a protective coating for steel (*galvanising*), and as a constituent of the important *brass* alloys. Common brass is about 1 part zinc to 2 parts copper, by weight.

Summary

Extraction

Metals are extracted from naturally occurring compounds known as minerals, present in ores.

Smelting consists of heating or roasting the ore, usually together with carbon as fuel (and reducing agent) and a flux, to drive off volatile impurities and reduce the oxidised metal. The flux lowers the melting point and combines with earthy impurities to form a slag.

In some cases, notably for aluminium, electrolytic methods of extraction are used.

Working

Metal products may be either wrought (by rolling, forging, pressing, drawing or extruding) or cast.

Jointing methods include welding, soldering (with soft solder) or brazing (with hard solder).

Testing

Standard tests on metals include the tensile test (giving yield stress, ultimate stress, percentage elongation, reduction in area and, Young's modulus (E)), the cold bend test, hardness (Brinell, Rockwell and Vickers), impact (Izod and Charpy) and examination of micro-structure.

Metals

Iron is smelted by kilning a mixture of ore, coke and limestone—the products are pig iron and blast-furnace slag. Pig iron contains up to 5% of carbon—present either as ferric carbide (in cementite), giving brittle white pig iron, or as graphite giving non-brittle grey pig iron.

Cast iron is obtained by re-smelting pig iron to obtain brittle white cast iron, or to obtain the less brittle and stronger, grey cast iron. Both have high compressive strength and are rust-resistant. White cast iron may be annealed to give malleable cast iron.

Wrought iron is refined pig iron containing less than 0·15% of carbon. It is ductile and corrosion-resistant.

Steel is an alloy of iron and carbon made by re-smelting pig iron, e.g. by the Siemens-Martin open hearth process. The metal is run off and moulded into ingots, or may be used directly to form cast steel products. Its carbon content can be controlled to give either low-carbon steels (0·1—0·3% carbon) or high carbon (tool) steels (1—2% carbon). In addition to the plain carbon steels there are special steel alloys, high-tensile steel, tough manganese steel, tungsten (tool) steel, stainless (non-rust) steel and spring (silicon) steel.

Aluminium is made by heating bauxite with caustic soda solution to dissolve the aluminium oxide, and then releasing the metal

K**

electrolytically using cryolite as flux and carbon electrodes. Aluminium is light in weight, highly resistant to corrosion, a good reflector and conductor and has high-strength alloys. It is easily rolled, drawn, extruded or cast.

Copper is obtained by roasting the sulphide copper pyrites and reducing it in a reverberatory furnace. It can be further refined by re-smelting, or by electrolytic method.

Copper is obtainable in soft, half-hard and hard temper, and is subject to work-hardening. It is highly resistant to corrosion, is a good conductor and is easily welded, soldered or brazed. It is a constituent of brass (with zinc) and bronze (with tin).

Lead is obtained by roasting and reducing the sulphide galena. The resulting pig lead is refined by re-smelting, or by electrolytic method. It is very malleable, is easily cut, soldered and welded, but has low tensile strength and creeps badly. Its ternary alloy, tellurium lead, is stronger and creeps much less. Lead is highly resistant to corrosion (except by alkalis).

Zinc is extracted from the sulphide ore zinc blende by roasting, which reduces the zinc oxide and condenses the metallic vapour formed. Zinc is light, easily worked and does not creep. It is highly resistant to corrosion, except electrolytically. It is a constituent of brass (with copper).

SELECTED DATA AND TEST REQUIREMENTS

TABLE I *Test requirements for cements, excluding chemical composition* (BS 12 : 1958)

BS TEST	SPECIFIED RESULT			
	Ordinary Portland		*Rapid-hardening*	
Fineness (min.)	2250 cm²/g		3250 cm²/g	
Compressive strength (min.) (av. of 3)	lbf/in²	(kgf/cm²)	lbf/in²	(kgf/cm)²
(a) At 3 days				
Mortar cubes	2200	(154)	3000	(210)
Concrete cubes	1200	(84)	1700	(119)
(b) At 7 days[1]				
Mortar cubes	3400	(239)	4000	(281)
Concrete cubes	2000	(140)	2500	(175)
Tensile strength (24 h) (optional)			lbf/in²	(kgf/cm²)
Av. of 6 (min.)			300	(21)
Setting time				
(a) Initial (min.)	45 min		45 min	
(b) Final (max.)	10 h		10 h	
Soundness (Le Chatelier) Expansion (max.)				
(a) Initial test[2]	10 mm (0·4 in)		10 mm (0·4 in)	
(b) Re-test[2]	5 mm (0·2 in)		5 mm (0·2 in)	

[1] The 7-day strengths must also exceed the 3-day strengths.
[2] If the first test shows a failure, the test is repeated on a sample exposed to air (R.H. 50—80%), in a 3 in thick layer for 7 days. The repeat test must also show a failure for the cement to be deemed unsound.

TABLE 2 *Test requirements for gypsum building plasters, excluding chemical composition* (BS 1191 : 1967 Part 1) (*See* Table 3 for premixed lightweight plasters)

TEST	SPECIFIED RESULT FOR PLASTER CLASS			
	A	B	C	D
Coarse particles: residue on BS 14 (1·20 mm) sieve not to exceed	5%	1% F, U/F	1% F	1% F
Transverse strength (oven-dried after 24 h): modulus of rupture (av. of 6) not less than	350 lbf/-in² (0·25 kgf/-mm²)	170 lbf/-in² (0·12 kgf/-mm²) U, U/F		
Mechanical resistance: (oven-dried after 24 h): impact indentation (av. of 16) diameter not to exceed		5 mm F, U/F	4·5 mm F	4·0 mm F
Expansion on setting not to exceed		0·20%[1] (24 hr)		
Soundness (6 pats)	Pats to show no sign of disintegration, popping or pitting			

[1]for board finish only
U—undercoat types
F—finishing types
U/F—dual purpose types

TABLE 3   *Test requirements for gypsum premixed lightweight plasters,*
excluding soluble salts and free lime content (BS 1191 : 1967 Part 2)

TEST	PLASTER TYPE				
	Browning	*Metal Lathing*	Bonding	*Multi-purpose*	*Final coat*
Dry bulk density not to exceed:					
lb/ft³	40	48	48	45	
(kg/m³)	(641)	(769)	(769)	(721)	
Dry set density not to exceed:					
lb/ft³	53	65	65	55	
(kg/m³)	(849)	(1041)	(1041)	(881)	
Compressive strength not less than:					
lbf/in²	135	145	145	145	
(gf/mm²)	(95)	(102)	(102)	(102)	
Mechanical resistance (oven-dried after 24 h): impact indentation dia. (av. of 16)					4·0 mm to 5·0 mm

TABLE 4 *Test requirements for hydrated lime powder and putty, excluding chemical composition* (BS 890 : 1966). *For requirements for quicklimes see* BS 890

TEST	TYPES			
	High calcium, semi-hydraulic and magnesian			
	Hydrated (powder)		Lime putty	
Fineness Residue on sieve not to exceed:				
(a) BS 85 (180 microns)	1%		1%	
(b) BS 170 (90 microns)[1]	6%		6%	
Soundness (a) Le Chatelier (av. of 3) Expansion not to exceed—	10 mm^2		10 mm^2	
(b) Pat test (3)	Free from pops or pits		Free from pops or pits	
Workability (flow table) Bumps to reach 190 mm Spread not less than	12		14	
Density of putty Not more than:				
(a) standard consistence	1·50 g/ml		1·45 g/ml	
(b) as supplied			1·50 g/ml	
Separation of water on standing Not more than			25 ml	
	Semi-hydraulic only			
Hydraulic strength Modulus of rupture at 28 days (av. of 6)	lbf/in^2	(gf/mm^2)	lbf/in^2	(gf/mm^2)
Not less than	100	(70)	100	(70)
Not more than	300	(210)	300	(210)

[1] Including residue on the BS 85 sieve
[2] Test not applicable to high calcium by-product limes.

Note The metric equivalents in this table are the author's (none is given in the British Standard). Values have been rounded, so are approximate only.

TABLE 5 *Minimum weight of sample for sieve analysis on aggregates* (BS 812 : 1967)

NOMINAL MAXIMUM SIZE		MINIMUM WEIGHT OF SAMPLE	
in	mm	lb	kg
$2\frac{1}{2}$	63·50	112	50
2	50·80	80	35
$1\frac{1}{2}$	38·10	35	15
$1\frac{1}{4}$	31·75	35	15
1	25·40	10	5
$\frac{3}{4}$	19·05	5	2
$\frac{1}{2}$	12·70	$2\frac{1}{2}$	1
$\frac{3}{8}$	9·52	1	0·5
$\frac{1}{4}$	6·35	$\frac{1}{2}$	0·2
$\frac{3}{16}$	4·76	$\frac{1}{2}$	0·2
Passing No. 7 mesh	2·40	$\frac{1}{4}$	0·1

Note BS 812 requires also that the balance used must be accurate to 0·5% of the weight of the test sample.

TABLE 6(a) *Metric equivalents of some* BS 410: 1962[1] *nominal sieve sizes* (*width of aperture*)

in	3	$1\frac{1}{2}$	1	$\frac{3}{4}$	$\frac{1}{2}$	$\frac{3}{8}$	$\frac{1}{4}$	$\frac{3}{16}$	$\frac{1}{8}$
mm	76·2	38·1	25·4	19·05	12·7	9·53	6·35	4·76	3·18
BS number	7	14	18	25	52	72	100	170	200
mm	2·40	1·20							
microns (μm)			850	600	300	210	150	90	75

[1] See Table 6(b) for ISO proposed and new BS 410:1969 nominal sieve sizes.

Note 1 micron=0·001 mm (It is known also as a micrometre, μm).

TABLE 7 *Maximum weight to be retained at the completion of sieving aggregates* (BS 812 : 1967)

BS sieve aperture width		Maximum weight				BS sieve			Maximum weight	
		18 in dia. sieves		12 in dia. sieves					8 in dia. sieves	
in	mm	lb	kg	lb	kg	Mesh No.	mm	microns	oz	g
2	50·80	22	10	10	4·5	7	2·40		7	200
1½	38·10	18	8	8	3·5	10	1·68		3½	100
1¼	31·75	13	6	5½	2·5	14	1·20		3½	100
1	25·40	13	6	5½	2·5					
						18		850	2½	75
¾	19·05	9	4	4½	2·0	25		600	2½	75
½	12·70	6½	3	3	1·5	36		420	2½	75
⅜	9·52	4½	2	2	1·0					
						52		300	2	50
¼	6·35	3	1·5	1½	0·75	72		210	2	50
3⁄16	4·76	2	1·0	1	0·5	100		150	1½	40
⅛	3·18			½	0·3	200		75	1	25

TABLE 8 *Container size and compaction for aggregate bulk density tests* (BS 812 : 1967)

CYLINDER						UN-COMPACTED DENSITY		COMPACTED DENSITY [1]		
Capacity		Diameter		Depth		Nom max. aggregate size		Nom max. aggregate size		No. of blows per layer [1]
ft³	litres	in	mm	in	mm	in	mm	in	mm	
1	28·3	14	355·6	11¼	285·2	2	50·80	2	50·80	100
½	14·2	10	254·0	11	279·4	½	12·70	1	25·40	50
¼	7·1	8	203·2	8½	215·9	¼	6·35	½	12·70	30
1⁄10	2·8	6	152·4	6	152·4			¼	6·35	20

[1] Container to be filled in 3 layers, each layer tamped by allowing a steel rod (⅝ in/16 mm) diameter and 24 in/600 mm long, rounded one end) to fall freely from a height of 2 in/50 mm above the surface of the aggregate.

TABLE 9   *Test requirement for clay, fine silt and fine dust in aggregates, as the amount passing* BS 200 *sieve* (*75 microns*)

	MAXIMUM TO PASS BS 200 SIEVE (PER CENT BY WEIGHT) FOR SAND IN:		
AGGREGATE TYPE	Concrete BS 882 : 1965	Granolithic Concrete BS 1201 : 1965	Plastering, rendering or brickwork BSS 1198, 1199, 1200 : 1955
Fine (passing $\frac{3}{16}$ in or 4·76 mm)			
(a) Natural or crushed gravel	3[1]	3[1]	2
(b) Crushed stone	15	8	10
Coarse	1	1	
All-in	Pro rata to above limits based on fine and coarse fractions		

[1] Corresponds approximately to 8% by volume using the field settling test of BS 812 : 1967
[2] The BS limit using the field settling test to BS 812 is 5%

TABLE 10   *Test requirements for aggregate impact value and* 10% *fines value tests for aggregates*

AGGREGATES IN	BS REF.	A.I.V. NOT TO EXCEED	10% FINES (MIN.) ton f (tonne f)
Normal dense concrete	882 : 1965	45%	5 (5·08)
Concrete for wearing surfaces	882 : 1965	30%	10 (10·16)
Granolithic concrete	1201 : 1965	—	15 (15·24)

Note The metric equivalents in this table are the author's (none is given in the British Standard)

TABLE 11 *Typical proportions for lime-gauged mortars for brickwork and blockwork*

LOCATION	CONDITION	MIX PROPORTIONS (BY VOLUME) CEMENT : LIME : SAND	HYDRAULIC LIME: SAND
External	Exposed or in autumn or winter	1 : 1 : 5–6[1]	1 : 2
	Sheltered or in Spring or summer	1 : 2 : 8–9[1]	1 : 3
Internal	Winter	1 : 1 : 5–6[1]	1 : 2
	Spring, summer or autumn:		
	(a) Normal density units	1 : 2 : 8–9[1]	1 : 3
	(b) Lightweight units	1 : 3 : 10–12[1]	1 : 3
Engineering Construction. Below ground d.p.c. level. Free-standing walls/parapets	All conditions	1 : 0–¼ : 3	

[1] The higher sand contents of the range given apply for well graded sands. In other cases (e.g. sands uniformly fine, or coarse) use a lower sand content.

Note See page 85 regarding alternative mortar mixes with plasticisers or masonry cement.
Note See also BS Codes of Practice 121:101 'Brickwork'.

TABLE 12   *Typical mortars for masonry*

JOINT	UNIT OR CONDITION	EXAMPLE	MIX PROPORTIONS (by volume)	
			Cement:lime: sand	*Hydraulic lime:sand*
	Dense	Granites, hard sandstones, hard limestones	1 : 0–$\frac{1}{4}$: 3 1 : 1 : 5–6[1]	
Thin	Low/medium density	Porous sandstones, porous limestones	1 : 3 : 10–12[1]	1 : 2–3[1]
	(a) Sheltered or summer		1 : 2 : 8–9[1]	
Normal		Rubble masonry		
	(b) Exposed or winter		1 : 1 : 5–6[1]	

[1] The higher sand contents of the range given apply for well graded sands. In other cases (e.g. sands uniformly fine or coarse) use a lower sand content.

Note See also BS Code of Practice 121:201 'Masonry walls'.

TABLE 13   *Grading requirements for sands in mortar* (BS 1200 : 1955)

BS SIEVE[1]	PERCENTAGE PASSING BS SIEVE	
	General purpose sands (plain brickwork, block- walling and masonry)	*Sands for reinforced brickwork*
$\frac{3}{16}$ in (4·76 mm)	100	100
No. 7 (2·40 mm)	90–100	90–100
No. 14 (1·20 mm)	70–100	70–100
No. 25 (600 μm)	40–100	40–80
No. 52 (300 μm)	5–70	5–40
No. 100 (150 μm)	0–15	0–10

[1] Metric equivalents are from BS 410 : 1962 'Specification for test sieves'—see Table 6(b) for proposed new nominal sieve sizes.

Note A tolerance of 5% (total) is permitted on the above limits, excluding those for the $\frac{3}{16}$ in sieve.

TABLE 14 *Grading requirements for sands in external renderings and floor screeds* (BS 1199 : 1955)

BS SIEVE[1]	PERCENTAGE PASSING BS SIEVE
$\frac{3}{16}$ in (4·76 mm)	100
No. 7 (2·40 mm)	90–100
No. 14 (1·20 mm)	70–100
No. 25 (600 μm)	40–80
No. 52 (300 μm)	5–40
No. 100 (150 μm)	0–10[2]

[1] Metric equivalents are from BS 410 : 1962 'Specification for test sieves'—see Table 6(b) for proposed new nominal sieve sizes.
[2] Higher limit permissible for crushed stone sands by agreement between supplier and purchaser (or his representative).

Note A tolerance of 5% (total) is permitted on the above limits, excluding those for the $\frac{3}{16}$ in sieve.

TABLE 15 *Grading requirements for sands in internal plastering* (BSS 1198–99 : 1955)

BS SIEVE[1]	PERCENTAGE PASSING BS SIEVE		Finishing coats
	Undercoats		
	Type 1 sand (all plastering mixes)	Type 2 sand (gypsum mixes only)	(all plastering mixes)
$\frac{3}{16}$ in (4·76 mm)	100	100	
No. 7 (2·40 mm)	90–100	90–100	100
No. 14 (1·20 mm)	70–100	70–100	90–100
No. 25 (600 μm)	40–80	40–100	55–100
No. 52 (300 μm)	5–40	5–50	5–50
No. 100 (150 μm)	0–10[2]	0–10[2]	0–10[2]

[1] Metric equivalents are from BS 410 : 1962 'Specification for test sieves'—see Table 6(b) for proposed new nominal sieve sizes.
[2] Higher limit permissible for crushed stone stands, by agreement between supplier and purchaser (or his representative).

Note A tolerance of 5% (total) is permitted on the above limits, excluding those for the $\frac{3}{16}$ in sieve in the case of undercoats, or the No. 7 sieve for finishing coats.

TABLE 16 *An outline of mixes for internal plastering—proportions by volume (See also* BS *Codes of Practice* 211 *'Internal Plastering')*

Note This table excludes premixed and special thin wall plasters which are normally used 'neat' as supplied.

1 UNDERCOATS

Cement-based[1]	Gypsum-based	
Use only on rigid backgrounds Useful in damp conditions	Grades are available to suit most surfaces Not suitable in permanently damp conditions	
Cement : lime : sand[2]	Plaster : sand	Plaster : lime : sand
1 : 0–¼ : 3 very strong, for rigid backgrounds with good key only 1 : 1 : 5–6[3] strong, for normal backgrounds 1 : 2 : 8–9[3] weak, for porous or lightweight background, or on metal lathing (or hydraulic lime : sand, 1 : 2–3)	1 : 1–3[3] Use Class B plaster	1 : 3 : 9 Use Class B plaster for lightweight background or on lathing, where to receive a weak lime-based finish

2 FINAL COATS WITH GYPSUM

Neat plaster	Plaster : lime	Lime : plaster (with or without sand)
Class B moderately hard Class C hard Class D very hard	Use Class B or C plaster Lime reduces hardness, improves workability and offsets setting expansion 1 : 0–⅓	Use Class B plaster For porous, soft finish—plaster offsets lime shrinkage 1 : ¼–½ (with 0–1 part of sand)

3 FINAL COATS WITH CEMENT

Use only with cement-based undercoats of similar strength—for very hard, strong finish or in damp conditions:
Cement : lime : sand
1 : 0–¼ : 3 (very strong and hard)
1 : 1 : 5–6[3] (strong, hard)
1 : 2 : 8–9[3] (weak)

[1] Cement-based mixes with sand may contain a proprietary plasticiser aerator in place of lime; or a masonry cement may be used (usually with reduced sand content—see manufacturer's instructions).
[2] Cement-based mixes are not generally suitable for application to

very low suction, smooth surfaces (unless spatterdash or a bonding agent is used).

[3] The lower sand contents of the range given apply to strong undercoats on low suction backgrounds and where finer (e.g. BS 1198 type 2) or poorly graded sands are used.

Notes

1 Proportions of lime given in the table refer to lime putty. Where the dry hydrate is used the lime can be increased by half to give the same workability.
2 Mixes based on lime or cement will not bond satisfactorily to gypsum plasterboard.

TABLE 17 *Grading requirements for concreting fine aggregate* (BS 882, 1021:1965)

BS SIEVE	PERCENTAGE PASSING BS SIEVE					
	Concrete aggregate BS 882				Grano aggregate BS 1201	
	Zone 1	Zone 2	Zone 3	Zone 4	Zone 1	Zone 2
⅜ in (9·52 mm)	100	100	100	100	100	100
3/16 in (4·76 mm)	90–100	90–100	90–100	95–100	90–100	90–100
No. 7 (2·40 mm)	60–95	**75–100**	85–100	95–100	60–95	75–100
No. 14 (1·20 mm)	30–70	**55–90**	75–100	**90–100**	30–70	**55–90**
No. 25 (600 μm)	15–34	35–59	60–79	80–100	15–34	35–59
No. 52 (300 μm)	5–20	**8–30**	**12–40**	15–50	5–20	**8–30**
No. 100 (150 μm)	0–10[1]	**0–10**[1]	0–10[1]	0–15[1]	0–10[2]	0–10[2]

Table 6(b) gives the proposed new nominal sieve sizes.

[1] Permissible limit for crushed stone sands 20%.
[2] Permissible limit for crushed stone sands 15%.

Note A tolerance of 5% (total) is permitted on the above limits, for the bold figures only.

TABLE 18 *Grading requirements for concreting coarse aggregates* (BS 882, 1201 : 1965)

BS SIEVE	PERCENTAGE PASSING BS SIEVE								
	Concrete aggregate BS 882								Grano aggregate BS 1201
	Graded				Single sized				Graded
	1½–3/16 in 38–4·76 mm	¾–3/16 in 19–4·76 mm	½–3/16 in 13–4·76 mm	2½ in 64 mm	1½ in 38 mm	¾ in 19 mm	½ in 12·7 mm	3/8 in 9·5 mm	3/8 in 9·5 mm
3 in (76·2 mm)	100			100					
2½ in (63·5 mm)				85–100					
1½ in (38·1 mm)	95–100	100		0–30	85–100	100			
¾ in (19·05 mm)	30–70	95–100	100	0–5	0–20	85–100	100		
½ in (12·7 mm)			90–100				85–100	100	100
3/8 in (9·53 mm)	10–35	25–55	40–85		0–5	0–20		85–100	85–100
3/16 in (4·76 mm)	0–5	0–10	0–10			0–5	0–45	0–20	0–20
7 in (2·40 mm)							0–10	0–5	0–5

Table 6(b) gives proposed new nominal sieve sizes.

TABLE 19   *Grading requirements for concreting all-in aggregates* (BS 882, 1201 : 1965)

BS SIEVE[2]	PERCENTAGE PASSING BS SIEVE		
	Concrete aggregate BS 882		Grano aggregate BS 1201
	$1\frac{1}{2}$ in (38 mm) nom.	$\frac{3}{4}$ in (19 mm) nom.	
3 in (76·20 mm)	100		
$1\frac{1}{2}$ in (38·10 mm)	95–100	100	
$\frac{3}{4}$ in (19·05 mm)	45–75	95–100	
$\frac{1}{2}$ in (12·70 mm)			100
$\frac{3}{8}$ in (9·53 mm)			95–100
$\frac{3}{16}$ in (4·76 mm)	25–45	30–50	30–60
7 (2·40 mm)			20–50
14 (1·20 mm)			15–40
25 (600 μm)	8–30	10–35	10–30
52 (300 μm)			5–15
100 (150 μm)	0–6	0–6	0–5

[2] Table 6(b) gives proposed new nominal sieve sizes.

TABLE 20   *Typical strength requirements for dense concrete test cubes with ordinary Portland cement*

MIX PROPORTION BY VOLUME ($\frac{3}{8}$ in/19 mm max. aggregate)	MINIMUM CUBE STRENGTH							
	Structural class				Foundation class			
	7 days lbf/in²	kgf/cm²	28 days lbf/in²	kgf/cm²	7 days lbf/in²	kgf/cm²	28 days lbf/in²	kgf/cm²
1 : 1 : 2 nom. preliminary[1]	4000	(280)	6000	(420)				
works	3000	(210)	4500	(315)				
1 : 1½ : 3 nom. preliminary[1]	3350	(235)	5000	(350)				
works	2500	(175)	3750	(265)				
1 : 2 : 4 nom. preliminary[1]	2700	(190)	4000	(280)				
works	2000	(140)	3000	(210)	1300	(90)	2000	(140)
1 : 3 : 6 nom. works					1000	(70)	1500	(105)
1 : 5 all-in works					1300	(90)	2000	(140)
1 : 7 all-in works					1000	(70)	1500	(105)

[1] Preliminary strengths are for laboratory trial mixes.

Note Metric equivalents in this table have been rounded and are therefore only approximate.

To convert from kgf/cm² to N/mm² divide by ten.

TABLE 21(a) Standard concrete mixes to BSCP 114 Part 1: 1957 (See TABLE 21(b) for metric data)

Specified works cube strength at 28 days	Weight of dry sand per 112 lb of cement	Weight of dry coarse aggregate per 112 lb of cement (Ordinary Portland or Portland blast furnace)											
		3/8 in (9·5 mm) maximum size			1/2 in (12·7 mm) maximum size			3/4 in (19 mm) maximum size			1½ in (38 mm) maximum size		
Workability		Low	Medium	High	Low	Medium	High	Low	Medium	High	Low	Medium	High
Slump (in)		0-¾	¼-1	1-2	¼-¾	¾-1½	1¾-4	½-1	1-2	2-5	1-2	2-4	4-7
Compacting factor		·80-·86	·86-·92	·92-·97	·81-·87	·87-·93	·93-·97	·82-·88	·88-·94	·94-·97	·82-·88	·88-·94	·94-·97
lbf/in^2	lb	lb	lb	lb	lb	lb	lb	lb	lb	lb	lb	lb	lb
3000	200	325	250	200	375	300	250	425	350	300	500	425	375
3750	175	275	200	150	325	250	200	375	300	250	450	375	325
4500	150	225	—	—	275	200	—	325	250	200	375	300	250

Notes

1 Above data assumes a standard deviation of 1000 lbf/in^2 (BSCP 116 gives additional mix data for use in precast works with 500 lbf/in^2 standard deviation).

2 Adjust quantities pro-rata for aggregates with specific gravity significantly different from 2·6.

3 Aggregates to comply with BS 882 (or BS 1047 for air-cooled blast-furnace slag). Single-sized aggregates to be combined accordingly.

4 Data given are based on BS 882 Zone 2 sand.
For Zone 1 sands increase the sand by at least 25 lb and reduce coarse aggregate by the same amount.
For Zone 3 sands decrease the sand by at least 25 lb and increase coarse aggregate by the same amount.
The data is not applicable to Zone 4 sands, nor to air-entrained concretes.
For crushed stone or crushed gravel sands reduce the coarse aggregate by at least 25 lb without altering the weight of sand.

5 Selection of mix proportions should take account of durability requirements where these indicate a richer mix than for strength alone.

6 If strength results are low use the next higher strength mix in the table (if none exists, increase the cement content by 10%).

7 A statistical analysis of 40 cube tests allows a final adjustment of the mix to be made.

TABLE 21(b) *Standard concrete mixes to* BSCP 114 Part 2: 1969 (see TABLE 21(a) *for imperial data*)

Specified works cube strength at 28 days	Weight of dry sand per 50 kg of cement	Weight of dry coarse aggregate per 50 kg of cement (ordinary Portland or Portland blast furnace)											
		10 mm maximum size			13 mm maximum size			19 mm maximum size			38 mm maximum size		
Workability		Low	Medium	High	Low	Medium	High	Low	Medium	High	Low	Medium	High
Slump (mm)		0–5	5–25	25–50	5–20	20–40	40–100	12–25	25–50	50–125	25–50	50–100	100–175
Compacting factor		·80–·86	·86–·92	·92–·97	·81–·87	·87–·93	·93–·97	·82–·88	·88–·94	·94–·97	·82–·88	·88–·94	·94–·97
N/mm^2	kg	kg	kg	kg	kg	kg	kg	kg	kg	kg	kg	kg	kg
21	90	145	110	90	165	135	110	190	155	135	225	190	165
25·5	80	125	90	65	145	110	90	165	135	110	200	165	145
30	65	100	—	—	125	90	—	145	110	90	165	135	110

Notes

1 Above data assumes a standard deviation of 7 N/mm^2 (BSCP 116 gives additional mix data for use in *precast works* with 3·5 N/mm^2 standard deviation).

2 Adjust quantities pro-rata for aggregates with specific gravity significantly different from 2·6.

3 Aggregates to comply with BS 882 (or BS 1047 for air-cooled blast-furnace slag). Single-sized aggregates to be combined accordingly.

4 Data given are based on BS 882 Zone 2 sand.
  For *Zone 1* sands increase the sand by at least 10 kg and reduce coarse aggregate by the same amount.
  For *Zone 3* sands decrease the sand by at least 10 kg and increase coarse aggregate by the same amount.
  The data is not applicable to Zone 4 sands, nor to air-entrained concretes.
  For crushed stone or crushed gravel sands reduce the coarse aggregate by at least 10 kg without altering the weight of sand.

5 Selection of mix proportions should take account of durability requirements where these indicate a richer mix than for strength alone.

6 If strength results are low use the next higher strength mix in the table (if none exists, increase the cement content by 10%).

7 A statistical analysis of 40 cube tests allows a final adjustment of the mix to be made.

TABLE 22   *Requirements for concrete roofing tiles* (BS 473, 550 : 1967)

REQUIREMENTS	DOUBLE LAP (GROUP A)		SINGLE LAP (GROUP B)
	Standard plain tiles	*Others*	
Nominal dimensions		Typical sizes[1] only are given	Typical sizes[2] only are given
	in mm	in mm	in mm
Length	$10\frac{1}{2}$ (267)		
Width	$6\frac{1}{2}$ (165)		
Thickness (min.)			
body	$\frac{3}{8}$ (9·53)	$\frac{3}{8}$ (9·53)	$\frac{3}{8}$ (9·53)
at sidelock			$\frac{1}{4}$ (6·35)
side lap (min.)			I (25·4)
Transverse strength (average of 6)[3] wet not less than: for test span of			
$7\frac{1}{2}$ in (190·5 mm)[6]	110 lbf (50 kgf)[6]		
$\frac{2}{3}$ tile length		12 × width in inches (lbf)	18 × effective width[4] in inches (lbf)
		(0·214 × width in mm for kgf)[6]	(0·321 × width in mm for kgf)[6]
Permeability (average of 3)[5] at 200 mm head, not to exceed, in cm³/min.:			
per in² of test area	0·00126	0·00126	0·00063
(per cm² of test area)	(0·000195)[6]	(0·000195)[6]	(0·000098)[6]

[1] These are 18 × 12 in (457 × 305 mm), 18 × 13 in (457 × 330 mm), 18 × 18 in (457 × 457 mm), and 28 × 18 in (711 × 457 mm). Other sizes may be agreed.

[2] These are 15 × 9 in (318 × 229 mm), $16\frac{1}{4}$/$16\frac{1}{2}$ × 13 in (413/419 × 330 mm), and 17 × 15 in (432 × 381 mm). Other sizes may be agreed.

[3] Or of 12 tiles if the average of 6 shows a failure.

[4] Effective width = width overall minus sidelock.

[5] Or of 6 tiles if the average of 3 shows a failure.

[6] Metric equivalents are by the author (none is given by the British Standards).

TABLE 23(a)   *Dimensions of clay bricks and blocks* (BS 3921 Part I : 1965)
(*excluding half and three-quarter length units, and special shapes*)
see TABLE 23(b) *for metric data*

Measurement	Dimensions (*in*)		Tolerance (*in²*)
	Nominal	Actual	
Bricks			On 24 bricks:
Length	9	$8\frac{5}{8}$	207 ± 3
Width	$4\frac{1}{2}$	$4\frac{1}{8}$	$99 \pm 1\frac{3}{4}$
Height	3	$2\frac{5}{8}$	$63 \pm 1\frac{3}{4}$
Blocks for walls			On each block (10 tested)[1]
Length	12	$11\frac{5}{8}$	$\pm \frac{3}{16}$
Width	$2\frac{1}{2}$, 3, 4	$2\frac{1}{2}$, 3, 4	$\pm \frac{3}{16}$
	6	6	$\pm \frac{1}{8}$
Height	9	$8\frac{5}{8}$	$\pm \frac{1}{8}$
Hollow blocks for structural floors and roofs			On each block (10 tested)[1]
Length	12	$11\frac{13}{16}$	$\pm \frac{3}{16}$
Width	12	$11\frac{13}{16}$	$\pm \frac{3}{16}$
Depth	3, 4	3, 4	$\pm \frac{3}{16}$
	5, 6, 7, 8, 9	5, 6, 7, 8, 9	$\pm \frac{1}{8}$
	10	10	$\pm \frac{3}{16}$

[1] If 3 or more results fail, the batch sampled does not comply with the BS. If 1, or 2, results fail, a further 10 shall be tested, of which failure of any one indicates non-compliance with the BS.

TABLE 23(b) *Dimensions of clay bricks and blocks* (BS 3921 Part 2:1969)
—*see* TABLE 23(a) *for imperial data*

Measurement	Dimensions (mm)		Tolerance (mm)
	Nominal	Actual	
Bricks			On 24 bricks:
Length	225	215	5160 ± 75
Width	112·5	102·5	2460 ± 45
Height	75	65	1560 + 60
			− 30
Blocks for walls			On each block (10 tested)[1]
Length	300	290	± 5·0
Width	62·5, 75, 100	62·5, 75, 100	± 2·5
	150	150	± 3·0
Height	225	215	± 3·0
Hollow blocks for structural floors and roofs			On each block (10 tested)[1]
Length	300	295	± 5·0
Width	300	295	± 5·0
Depth	75, 100	75, 100	± 2·5
	125, 150, 175	125, 150, 175	± 3·0
	200, 225	200, 225	± 3·0
	250	250	± 5·0

[1] If 3 or more results fail, the batch sampled does not comply with the BS. If 1, or 2, results fail, a further 10 shall be tested, of which failure of any one indicates non-compliance with the BS.

TABLE 24   *Test requirements for clay engineering bricks* (BS 3921 : 1965/1969)—*see Table 23 for dimensions*

TEST	RESULT	
	Class A	Class B
Compressive strength		
Average of 10 not less than	10,000 lbf/in²	7.000 lbf/in²
	69 N/mm²	48·5 N/mm²
Water absorption (5 h boiling)		
Average not to exceed	4·5%	7·0%

Note The metric data in this table comply with BS 3921 Part 2: 1969.

TABLE 25 *Compressive strength of clay bricks and blocks*
(BS 3921 : 1965/1969)—*see* TABLE 24 *for data on engineering bricks*

Bricks—lowest strength for (average of 10)	lbf/in²	N/mm²
(a) loadbearings walls	750	5·2
(b) non-loadbearing walls	200	1·4

Class No[1]	1	2	3	4	5	7	10	15
Compressive strength Average of 10 not less than:								
lbf/in²	1000	2000	3000	4000	5000	7000	10,000	15,000
N/mm²	7·0	14·0	20·5	27·5	34·5	48·5	69·0	103·5

Blocks for walling (average of 10)	lbf/in²	N/mm²
lowest strength for (a) loadbearing walls	400	2·8
(b) non-bearing walls	200	1·4

Blocks for structural floors and roofs (average of 10) strengths to be not less than	2000	14·0

[1] Intermediate class numbers may be interpolated (e.g. class 2·5 indicates 2500 lbf/in² or 17·25 N/mm²).

Note The metric data in this table comply with BS 3912 Part 2: 1969.

TABLE 26 *Test requirements for soluble salts content[1] of clay bricks and blocks for walling* (BS 3921 : 1965/1969)

Soluble salt radicle	BS limit (*max.*) as per cent by weight for	
	Ordinary quality and units for internal walls	Special quality
Sulphate		0·30
Calcium		0·10
Magnesium	Not specified	0·03
Potassium		0·03
Sodium		0·03

[1] Determined by the method of BS 3921.

TABLE 27 *Efflorescence of clay bricks and blocks—classification for tests to* BS 3921 : 1965/1969

Classification	Description
Nil	No perceptible deposit of efflorescence
Slight	Not more than 10% of the area of the face covered with a thin deposit of salts
Moderate	A heavier deposit than 'slight' and covering up to 50% of the area of the face but unaccompanied by powdering or flaking of the surface
Heavy	A heavy deposit of salts covering 50% or more of the area of the face, but unaccompanied by powdering or flaking of the surface
Serious	A heavy deposit of salts accompanied by powdering and/or flaking of the surface and tending to increase with repeated wettings of the specimen

Note BS 3921 : 1965/1969 specifies that 10 samples are tested—no sample brick or block for *walling* is to develop efflorescence worse than moderate (no requirement is given for blocks for structural floors and roofs).

TABLE 28 *Test requirements for calcium silicate (sandlime and flintline bricks)* (BS 187 : 1967/1970)—*see Table 29 for dimensions*

TEST	BRICK CLASS							
	I	2A	2B	3A	3B	4	5	
Crushing strength (wet) Average of 10 not less than								
lbf/in^2	1000	2000	2000	3000	3000	4000	5000	7000
N/mm^2	7·0	14·0	14·0	20·5	20·5	27·5	34·5	48·5
Coefficient of variation[1] (%) not to exceed	30	30	30	20	20	20	16	16
Drying shrinkage (%) Average of 3 not to exceed	—	0·025	0·035	0·025	0·035	0·025	0·025	0·025

[1] A statistical measure of dispersion (see BS 187).

TABLE 29 *Dimension of calcium silicate (sandlime and flintlime) bricks* Part I

BS 187 Part I : 1967 imperial units			BS 187 Part 2 : 1970 metric units		
LENGTH	WIDTH	HEIGHT*	LENGTH	WIDTH	HEIGHT*
in	in	in	mm	mm	mm
$8\frac{5}{8}\begin{smallmatrix}+\frac{1}{16}\\-\frac{1}{8}\end{smallmatrix}$	$4\frac{1}{8}\pm\frac{1}{16}$	$2\frac{5}{8}\pm\frac{1}{16}$	$215\begin{smallmatrix}+2\\-3\end{smallmatrix}$	103 ± 2	65 ± 2

* Bricks of height $2\frac{7}{8}$ in or 73 mm that otherwise comply with this BS may be available in certain areas.

TABLE 30 *Dimensions of concrete bricks* (BS 1180 : 1944)

LENGTH	WIDTH	HEIGHT¹
in	in	in
$8\frac{3}{8}\pm\frac{1}{8}$	$4\frac{1}{8}\pm\frac{1}{16}$	$2\frac{5}{8}\pm\frac{1}{16}$

[1] Bricks of $2\frac{7}{8}$ in height are omitted from this table as this is no longer a standard size.

Note To convert inches to millimetres, multiply by 25·4.

TABLE 31 *Test requirements for concrete bricks, excluding dimensions* (BS 1180 : 1944)

TEST	BRICK CLASS			
	Special purposes	A (i)	A (ii)	B
Compressive strength Average of 12 not less than				
lbf/in²	2500	1750	1750	1000
kgf/cm²	(176)	(123)	(123)	(70)
Drying shrinkage (%) Average of 3 not to exceed	0·025	0·025	0·04	0·06

Note The metric equivalents in this table are the author's (none is given in the British Standard). Values have been rounded, and are therefore only approximate.

TABLE 32 *Test requirements for clay plain roofing tiles size $10\frac{1}{2} \times 6\frac{1}{2}$ in (265×165 mm) and $\frac{3}{8}$–$\frac{5}{8}$ in (10–15 mm) thick (BS 402 : 1945/1970)*

Test	Result
Transverse strength (wet) Av. of 6[1] (at $7\frac{1}{2}$-in/180 mm span) not less than	175 lbf (778 N)
Water absorption Av. of 6[2] not to exceed	10·5%

[1] Or of 12 tiles if the average of six shows a failure.
[2] If the first test fails a repeat test is made and must show a pass.

Note The metric data in this table comply with BS 402 Part 2 : 1970.

TABLE 33   *Simplified key* for the microscopical identification of common softwoods (see also 'Notes on Softwood Key' page 322)*
I Resin Ducts (Canals) normally present (Ray tracheids[2] normally also present)

A	B	C
Horizontal ducts[1] CIRCULAR *Thin-walled epithelial[4] cells, duct extending to edge of ray* PINUS spp	*Horizontal ducts[1]* OVAL *Thick-walled epithelial[4] cells* 7–12	*Horizontal ducts[1]* ANGULAR *Very thick-walled epithelial[4] cells* 5–6
(a) Ray tracheids[2] *Dentate* Window-like pits 1–2[8] per crossfield. Pinus sylvestris (Scots pine. European redwood)	(a) Epithelial cells[4] *mostly more than* 9. Resin ducts towards one end of ray. Pits small (mainly piceoid)[8] in horizontal lines. Occasional spiral thickening[9] *Larix* spp (*Larch*) Laris decidua (European larch)	Abundant spiral thickening.[9] Ray tracheids[2] often shallow. Pseudotsuga taxifolia*(Douglas fir)
(b) Ray tracheids[2] *Dentate[5] and Reticulate[6]* Pits pinoid[8] 3–6 per crossfield. Pinus palustris and pinus echinata (Pitch pine)	(b) Epithelial[4] cells *mostly less than* 9. Pits small (piceoid)[8] in corners of crossfield[7]. Occasional spiral thickening. *Picea* spp (*Spruces*)	* Pseudotsuga menziesii is proposed for the next revision of BS 881
(c) Ray tracheids[2] *Smooth-walled* Pits angular window-like[8] 1–3 per crossfield[7] (i) Fusiform[3] rays dumpy. Pinus strobus (Yellow pine. Northern white pine) (ii) Fusiform[3] rays fairly high. Pinus monticola (Western white pine)	(i) Ray cells in tangential section *elongated*. Picea abies (European whitewood). Picea glauca (Canadian spruce) (ii) Ray cells in tangential section *Squarish* Picea sitchensis (Sitka spruce)	

* Abridged version of Timber Research and Development Association publication 'Timber Information' Ref. No 25 'Key to the microscopical identification of the principal softwoods used in Great Britain'—to which readers are referred for full information.

L

II *Resin Ducts (Canals) normally absent*

II RAY TRACHEIDS[2] NORMALLY PRESENT	RAY TRACHEIDS NORMALLY ABSENT	
(a) Bordered pits in tracheids with *scalloped tori*[10] *Cedrus* spp (*true cedars*) e.g. Cedrus libani (Cedar of Lebanon)	Bordered pits when more than uniseriate[13] always *opposite* (a) Tracheids with *loose spiral thickening*[9] Taxus baccata (*Y*ew) (b) Tracheids *without spiral thickening*	Bordered pits in tracheids *alternate*[15] Leaf traces[16] often present Araucaria angustifolia (Parana pine)
(b) Bordered pits in tracheids without *scalloped tori* Tsuga spp (*Hemlocks*) Ray cells often containing granular resin[11] seen in a crossfield (i) Longitudinal resin cells[12] frequent Tsuga heterophylla (Western hemlock) (ii) Longitudinal resin cells[12] absent Tsuga canadensis (Eastern hemlock)	I Horizontal walls of ray cells well pitted (taxodioid)[8] *Abies* spp Abies alba (Silver fir) II Horizontal walls of ray cells thick and sparsely pitted (i) end walls of ray cells nodular[14] Juniperus virginiana (Virginian pencil cedar) (ii) end walls of ray cells smooth horizontal walls thin. Small crossfield pits (taxodioid[8] common) Podocarpus spp (Podo) (iii) pencil odour, abundant parenchyma Juniperus procera (African pencil cedar) (iv) parenchyma usually scarce— Thuja plicata (Western red cedar)	

Notes on Softwood Key

1 *Horizontal resin duct.* A resin duct carried horizontally through a ray

2 *Ray tracheid.* A tracheid lying horizontally along a ray

3 *Fusiform ray.* One carrying a horizontal resin duct

4 *Epithelial cells.* Parenchyma cells lining a resin duct (encircling its section)

5 *Dentate.* Toothed or jagged outline

6 *Reticulate.* Appearing almost interlocked

7 *Crossfield.* Section through a parenchyma ray cell at its junction with a vertical tracheid

8 *Pits* (here referring to crossfield pitting):
   (a) *Window-like.* Large and simple without border ▭
   (b) *Pinoid.* Fairly small, irregular in shape and size. Simple, or with narrow border ◊ ○
   (c) *Piceoid.* With slit-like aperture, often extending beyond the pit border ⊘
   (d) *Cupressoid.* Aperture smaller than the pit border (usually oval) ⊘
   (e) *Taxodioid.* Aperture oval to circular, wider than the pit border ⊘

9 *Spiral thickening.* Seen as helical lines in tracheid walls

10 *Scalloped tori.* Where the wall surrounding the torous pad of a bordered pit is serrated

11 *Granular resin.* Usually seen as crystals with sharply defined edges

12 *Resin cell.* A tracheid blocked with resin

13 *Uniseriate pitting.* Pits in a single, vertical line down the tracheid wall

14 *Nodular cell walls.* Ray cells showing bulbous thickenings at their end walls

15 *Pits alternate.* Pits staggered alternately so that adjacent vertical lines appear to interlock

16 *Leaf trace.* A small, roundish, dark spot like a tiny knot in appearance and the size of a leaf-stalk

TABLE 34   *Grading of plywood manufactured from tropical hardwoods* (BS 1455 : 1963)

Plywood shall be graded according to the appearance of the face and back, each being assessed separately after the board has been made and not when in the form of veneer as defined in Clause 2.[1]

Grades of veneer are defined as follows:

Grade 1 veneer. Grade 1 veneer shall be of one or two pieces of firm, smoothly cut veneer. When of two pieces the joint shall be approximately at the centre of the board. The veneers shall be reasonably matched for colour. The veneer shall be free from knots, worm and beetle holes, splits, dote, glue stains, filling or inlaying of any kind or other defects. No end joints are permissible.

Grade 2 veneer. Grade 2 veneer shall present a solid surface free from open defects. When jointed, veneers need not necessarily be matched for colour or be of equal width. A few sound knots are permissible, with occasional minor discoloration and slight glue stains and isolated pinholes not along the plane of the veneer. Occasional splits not wider than $\frac{1}{16}$ in (0·8 mm) at any point and not longer than one-tenth of the length of the panel or slightly opened joints may be filled with a suitable filler. This grade shall admit neatly made repairs consisting of inserts of the same species as the veneer, which present solid, level, hard surfaces and are bonded with an adhesive equivalent to that used for bonding the veneers. No end joints are permissible.

Note   Grade 2 veneer excluding pinholes can be supplied by agreement between purchaser and supplier.

Grade 3 veneer. Grade 3 veneer may include wood defects, including worm-holes, which are excluded from Grades 1 and 2 in number and size which do not impair the serviceability of the plywood. It may also include manufacturing defects, such as rough cutting, overlaps, gaps or splits, provided these do not affect the use of the plywood. No end joints are permissible.

Other grades. Other grades, appropriate to the end use, may be agreed between purchaser and supplier.

[1] See BS 1455 : 1963.

TABLE 35 *Bonding of plywood made from tropical hardwoods* (BS 1455 : 1963)

Bonding between veneers shall be WBP, BR, MR or INT defined[1] as follows, and these designatory letters shall be used in marking the plywood.

Type WBP: Weather and boil-proof. Adhesives of the type[2] which by systematic tests and by their records in service over many years have been proved to make joints highly resistant to weather, micro-organisms, cold and boiling water, steam and dry heat.

Type BR: Boil resistant. Joints made with these adhesives have good resistance to weather and to the boiling water test, but fail under the very prolonged exposure to weather that Type WBP adhesives will survive. The joints will withstand cold water for many years and are highly resistant to attack by micro-organisms.

Type MR: Moisture-resistant and moderately weather-resistant. Joints made with these adhesives will survive full exposure to weather for only a few years. They will withstand cold water for a long period and hot water for a limited time, but fail under the boiling water test. They are resistant to attack by micro-organisms.

Type INT: Interior. Joints made with these adhesives are resistant to cold water but are not required to withstand attack by micro-organisms.

[1] These designations are those used in BS 1203, 'Synthetic resin adhesives (phenolic and animoplastic) for plywood'.
[2] At present only certain phenolic adhesives have been shown to meet this requirement.

TABLE 36 *Classification of glass for glazing* (BS 952 : 1964)

Sheet glass	Clear plate glass
(i) Ordinary glazing quality (O.Q.): for general glazing purposes, e.g. for factories, housing estates, etc.	(i) Glazing quality for glazing (G.G.): standard for general glazing purposes.
(ii) Selected glazing quality (S.Q.): for glazing work requiring a selected sheet glass above the ordinary glazing quality.	(ii) Selected glazing quality (S.G.): for better-class work, and also for mirrors and bevelling.
(iii) Special selected quality (S.S.Q.): for high-grade work where a superfine sheet glass is required, e.g., pictures, cabinet work, etc.	(iii) Silvering quality (S.Q.): for high-grade mirrors, and for purposes where a superfine glass is required.
(iv) Horticultural quality: an inferior quality, available in limited sizes for horticultural purposes.	

Note Selection for qualities of surface and metal is made to obtain glasses for qualities (ii) and (iii) but it must be realised that distortion cannot be completely eliminated from sheet glass.

CONVERSION FACTORS

TO

←————————→

	Multiply by			*Multiply by*	
Length	0·0394	in	mm	25·4	Length
	3·2808	ft	m	0·3048	
	1·0936	yd	m	0·9144	
	0·6214	mile	km	1·6093	
Area	0·155	in²	cm²	6·452	Area
	10·764	ft²	m²	0·0929	
	1·196	yd²	m²	0·8361	
Volume	0·061	in³	cm³ (ml)	16·39	Volume
	35·315	ft³	m³	0·0283	
	1·308	yd³	m³	0·7646	
Capacity	0·22	gal (UK)	litre (dm³)	4·546	Capacity
	0·0353	ft³	dm³	28·3	
Mass	2·2046	lb	kg	0·4536	Mass
	0·9842	ton (UK)	tonne (Mg)	1·0161	
Density	62·43	lb/ft³	g/cm² (g/ml)[1]	0·016	Density
	0·0624	lb/ft³	kg/ m³	16·02	
Force	2·2046	lbf	kgf	0·4536	Force
	0·2248	lbf	N	4·448	
	0·9842	tonf (UK)	tonnef	1·0161	
	0·1003	tonf (UK)	kN	9·964	
	0·102	kgf	N	9·807	
Stress and Pressure	14·223	lbf/in²	kgf/cm²	0·0703	Stress and Pressure
	145	lbf/in²	N/mm²	0·0069	
	0·2048	lbf/ft²	kgf/in²	4·882	
	0·02089	lbf/ft²	N/m²	47·88	
	0·635	tonf/in² (UK)	kgf/mm²	1·575	
	0·06475	tonf/in² (UK)	N/mm²	15·44	
	0·914	tonf/ft² (UK)	kgf/mm²	1·094	
	0·0914	tonf/ft² (UK)	tonnef/m²	10·94	
	10·2	kgf/cm²	N/mm²	0·09807	
	10	bar	N/mm²	0·1	
	0·98	bar	kgf/cm²	1·02	
	1·54	hectobar	tonf/in²	0·648	

[1] 1 g/cm³ (g/ml) = 1 kg/dm³ (kg/l) = 1 tonne/m³.

[2] 1 kg/m³ = 1 g/dm³ = 1 mg/cm³.

330

COMMON METRIC UNIT MULTIPLES
AND SUB-MULTIPLES

Symbol	Prefix	Multiplication factor	
M	mega	1,000,000	(one million or 10^6)
k	kilo	1000	(one thousand or 10^3)
h	hecto	100	(one hundred or 10^2)
da	deca	10	(ten or 10^1)
d	deci	0·1	(one-tenth or 10^{-1})
c	centi	0·01	(one hundredth or 10^{-2})
m	milli	0·001	(one thousandth or 10^{-3})
μ	micro	0·000,001	(one millionth or 10^{-6})

SOME BRITISH STANDARDS
AND CODES OF PRACTICE[1]

Note British Standards and Codes of Practice are amended and revised periodically.

BRITISH STANDARDS

12 : 1958	Portland cement (Ordinary and rapid-hardening)
146 : 1958	Portland blast-furnace cement
187 : 1967, 1970*	Specification for calcium silicate (sandlime and flintlime) bricks
373 : 1957	Methods of testing small clear specimens of timber
402 : 1945, 1970*	Clay plain roofing tiles and fittings
410 : 1969	Specification for test sieves (metric units)
473, 550 : 1967	Specification for concrete roofing tiles and fittings
556 : 1966	Specification for concrete cylindrical pipes and fittings including manholes, inspection chambers and street gullies
598 : 1958	Sampling and examination of bituminous mixtures for roads and buildings
690 : 1963	Asbestos-cement slates, corrugated sheets and semi-compressed flat sheets
743 : 1966, 1970*	Specification for materials for damp-proof courses
745 : 1969	Animal glues for wood
747 : 1968	Specification for roofing felts
812 : 1967	Methods for sampling and testing of mineral aggregates, sands and fillers
881, 589 : 1955	Nomenclature of commercial timbers including sources of supply
882, 1201 : 1965	Specification for aggregates from natural sources for concrete (including granolithic)
890 : 1966	Specification for building limes
915 : 1947	High alumina cement
952 : 1964	Classification of glass for glazing and terminology for work on glass

[1] Copies of publications in this list may be purchased from: British Standards Institution, Sales Branch, 101-113 Pentonville Road, London, N.1.

1047 : 1952	Air-cooled blast-furnace slag coarse aggregate for concrete
1053 : 1966	Specification for water paint and distemper for interior use
1142 : 1961	Specification for fibre building boards
1180 : 1944	Concrete bricks and fixing bricks
1186 (Part 1): 1952	Quality of timber in joinery
1191 (Part 1): 1967	Specification for gypsum building plasters—excluding premixed lightweight plasters
1191 (Part 2): 1967	Specification for gypsum building plasters—premixed lightweight plasters
1194: 1955, 1969*,	Concrete porous pipes for under-draniagde
1198, 1199, 1200: 1955	Building sands from natural sources
1201, 882 : 1965	Specification for aggregates from natural sources for concrete
1203 : 1963	Specification for synthetic resin adhesives (phenolic and aminoplastic) for plywood
1204 : 1964	Specification for synthetic resin adhesives. Gap filling (phenolic and aminoplastic) for constructional work in wood
1281 : 1966	Specification for glazed ceramic tiles and tile fittings for internal walls
1286 : 1945	Clay tiles for flooring (dimensions and workmanship only)
1370 : 1958	Low heat Portland cement
1391 : 1952	Performance tests for protective schemes uscd in the protection of light-gauge steel and wrought iron against corrosion
1444 : 1970	Cold-setting casein glue for wood
1455 : 1963	Specification for plywood manufactured from tropical hardwoods
1860 (Part 1): 1959	Structural timber, measurement of characteristics affecting strength
1881 : 1970	Methods of testing concrete (metric edition)
2028 : 1968	Precast concrete blocks (metric units)
2521/3 : 1966	Specifications for lead-based priming paints
2524 : 1966	Specification for red oxide-linseed oil priming paint
2525/7 : 1969	Specification for undercoating and finishing paints for protective purposes (white lead based) (metric units)
2604 : 1963	Resin-bonded wood chipboard
2660 : 1955	Colours for building and decorative paints
3235 : 1964	Test methods for bitumen

3260 : 1969 Specification for PVC (vinyl) asbestos floor tiles
3261 : 1960 Specification for flexible PVC flooring
3444 : 1961 Blockboard and laminboard
3544 : 1962 Methods of test for polyvinyl acetate adhesives for wood
3921 : 1965, 1969* Specification for bricks and blocks of fired brickearth, clay or shale
4036 : 1966 Specification for asbestos-cement fully compressed flat sheets
4071 : 1966 Specification for polyvinyl acetate (PVA) emulsion adhesives for wood
4551(Part 1):1970 Methods of testing mortars and specification for mortar testing sand (metric units)

BRITISH STANDARD CODES OF PRACTICE

112 : 1967, 1970* The structural use of timber
114 : 1957, 1969* The structural use of reinforced concrete in buildings
116 : 1965, 1969* The structural use of precast concrete
121 : 101 (1951) Brickwork
121 : 201 (1951) Masonry walls ashlared with natural stone or with cast stone
211 : 1966 Internal plastering

* Where two publication dates are given the first refers to Part 1 (imperial units) and the second to Part 2 (metric units). It is intended that the metric edition should supersede the imperial edition.

EXERCISES

1 Explain the following terms in relation to Portland cement (a) hydration (b) setting (c) hardening.
State the BS specified setting times for ordinary and rapid-hardening Portland cements. Comment on the fact that they are the same for both cements.

2 Say when you would expect the following types of cement to be used (a) sulphate resisting cement (b) extra rapid-hardening cement (c) high alumina cement. Give reasons.

3 State briefly the essential differences in the raw materials, manufacture and composition of *ordinary Portland* and *high alumina* cements. In what circumstances would you consider using high alumina cement in preference to ordinary Portland?

4 A batch of Portland cement is suspected of being stale. How would you check this?

5 Name the different British Standard tests applied to ordinary and rapid-hardening Portland cements.
Outline the method of any two BS tests for these cements, sketching the apparatus used.

1 What is the chemical name of the raw material used in the manufacture of gypsum plaster? What chemical difference is there between the raw material and the finished product?

2 Give the number and name of the British Standard which deals with gypsum plasters for building. State and name the different classes of plaster designated in the BS, and distinguish between them in respect of their manufacture, composition and properties.

335

3 Compare the setting and hardening action of a gypsum plaster with that of a Portland cement.

EXERCISES (CHAPTER 3)

1 Name the raw material from which building limes are manufactured. Explain the difference between the following types of lime, with reference to the raw material and the finished product. (a) non-hydraulic limes (b) hydraulic limes (c) magnesian limes.

2 List the British Standard tests applicable to building limes. Outline any two of the tests, giving a sketch of the apparatus used.

3 In what natural forms is the compound calcium carbonate found? Explain the following: (a) calcination of limestone (b) slaking of lime (c) carbonation of lime.

EXERCISES (CHAPTER 4)

1 Distinguish between 'fine', 'coarse' and 'all-in' aggregates. Explain what is meant by the 'grading' of an aggregate. Why is an aggregate's grading of importance?

2 Explain what is meant by the 'bulking' of sand. How is bulking related to the moisture content and grading of a sand?

3 Give one example where a highly porous aggregate would be desirable in concrete, and one example where it would be unsuitable for use in concrete. Name two types of aggregate suitable for use in concrete where a high degree of fire resistance is required.

4 State, in general terms, the relationship between the specific surface of an aggregate and its particle shape, surface texture and fineness modulus. Values of fineness modulus for a fine and a coarse aggregate are 2·80 and 7·00 respectively. Calculate (a) the fineness modulus for a combined grading with 40% fine aggregate by weight (b) the percentage of fine aggregate for a combined grading with a fineness modulus of 5·70.

EXERCISES (CHAPTER 5)

1 For what purpose is mortar used in building, and what types of mix are used?

2 What are the effects of (a) lime and (b) cement, in brickwork mortar?

3 Explain the difference between mortar and concrete.
List four main types of mortar mix used in brickwork. For what purpose would mortars of the following mix proportions (by volume) be suitable (a) 1:3 cement: sand (b) 1:1:6 cement:lime: sand (c) 1:2:9 cement:lime:sand?

4 Outline, with reasons, the desirable properties of mortars for brickwork.

EXERCISES (CHAPTER 6)

1 Explain the meaning of the following in relation to external rendering: (a) pebble dash (dry-dash) (b) roughcast (wet-dash) (c) tyrolean.

2 What measures may be adopted to secure an adequate bond or keying of external rendering applied to rigid backgrounds?

EXERCISES (CHAPTER 7)

1 Summarise the important differences between cement-based and gypsum-based internal plasterwork, indicating their advantages and disadvantages.

2 What are premixed plasters?
What advantages do premixed lightweight plasters have over other types?

EXERCISES (CHAPTER 8)

1 Explain the difference between the following, when applied to concrete: (a) nominal mixes and standard mixes (b) volume batching and weight batching.

2 Why is it important that structural dense concrete should be fully compacted?

3 Describe how you would make a concrete works test-cube.
State how the strength of hardened concrete is influenced by (a) the water:cement ratio (b) age (c) curing (d) the type of cement used.

4 A concrete mix is specified as $1:2\frac{1}{2}:5$ (cement:sand:gravel) by weight with a water:cement ratio of 0·50. The moisture content of the coarse aggregate is 1% and that of the sand 5%. How much water should be added at the mixer for a batch containing 200 kg of cement?

5 A concrete mix is specified to be of proportions 1:2:4 (cement: fine: coarse) by volume, with a water:cement ratio 0·60.
Batching is to be by weight and a mix with 180 kg of cement is to be used.
Determine the amounts of the fine and coarse aggregates, and water, to be measured per batch if the dry bulk densities of aggregates are 1·68 kg/dm³ (fine) and 1·52 kg/dm³ (coarse) (a) assuming dry aggregates are used (b) when using damp aggregates of moisture content 5% (fine) and 2% (coarse) by dry weight. (Take the cement bulk density as 1·43 kg/dm³.)

EXERCISES (CHAPTER 9)

1 For what types of work are clay engineering bricks used, and what two particular properties are they required to have?
Name the two important standard tests applied to engineering bricks and state any relevant BS requirements. Suggest why the the same requirements should not apply to clay facing bricks.

2 Which British Standard deals with requirements for calcium silicate (sandlime and flintlime) bricks?
Find out, and state, what method of classification is used in the British Standard for these bricks (tabulate the relevant information). What tests, other than for dimensions, may be applied for sandlime bricks to establish their conformity to a particular BS class? State the method of measuring sandlime bricks to determine their compliance with BS limits for dimensions differs from the method used for clay building bricks, and suggest a reason for this difference of method.

3 List the various stages in the manufacture of clay roofing tiles.

What points would you check in a preliminary field examination of a consignment of clay and concrete roofing tiles, in respect of their quality?

Name the principal BS tests applied to clay and concrete roofing tiles respectively. Comment on any difference in the tests applied.

4 Name four different kinds of ware used in the manufacture of sanitary goods.

Explain the difference between glazed earthenware and vitreous china. Describe the manufacture of fireclay.

5 What types of clay product are available for floor tiling? Describe the principal differences in their manufacture and properties.

EXERCISES (CHAPTER 10)

1 Name the three groups into which natural rocks are classified geologically.

Explain briefly the way in which the rocks in each of the three groups have been formed.

To which of these geological groups does each of the following belong: (a) sandstone (b) marble (c) granite?

2 What factors related to composition and physical properties determine the durability of (a) sandstones (b) limestones?

3 Give two examples of each of the following types of natural stone. Give general information as to their source, characteristics, properties and typical uses (a) sandstone (b) limestone (c) granite.

EXERCISES (CHAPTER 11)

1 Explain each of the following terms with reference to a neatly drawn section through the trunk of a tree: (a) heartwood (b) sapwood (c) pith (d) growth ring (e) springwood (f) summerwood (g) cambium (h) ray.

2 Draw up a table listing the main differences between softwoods and hardwoods.

3 State how moisture content affects the following properties of timber: (a) dimensions (b) strength (c) resilience (d) heat insulation.

Calculate the moisture content of a sample of timber from the following test results: initial weight of sample 25·0 g; oven-dry weight of same sample 20·0 g.

If the sample was cut from a plank taken from a kiln, and its weight on return to the kiln immediately after sampling was 15·0 kgf, what should be the plank's weight when the batch of timber is kilned to 10% moisture content?

4 Summarise the principal working characteristics, properties and uses of any four commercially important softwoods, and four hardwoods.

5 Explain the meaning of the terms 'ring-porous' and 'diffuse-porous' when applied to timber.
Which section of a timber sample would you examine to determine whether or not it is ring-porous?
Of the three timbers ash, pitch pine and beech, state which is (a) ring-porous (b) diffuse-porous (c) without pores.

6 It is said that dampness is the main cause of wood-rot in buildings. What other factors will assist the development of fungal attack? List the common causes of dampness in buildings.
Why is dry rot considered a more serious form of attack than wet rot?

7 How would you distinguish between an attack by the common dry rot (merulius lacrymans) and wet rot (coniophora cerebella) in a building?

EXERCISES (CHAPTER 12)

1 Describe the construction of each of the following, and state their advantages over solid timber (a) plywood (b) blockboard (c) laminboard.

2 Name the three main grades in which hardboard is manufactured. Indicate the distinctive properties and uses of these three grades of hardboard.
Find out, describe, and account for the procedure recommended for the conditioning of hardboard prior to fixing.

3 Describe the composition of asbestos-cement sheets, and name and distinguish between the different types of flat sheet produced.

4 Explain any special precautions which apply to the use of asbestos-

cement sheets in respect of (a) painting (b) plastering (c) nailing (d) the application of loads.

EXERCISES (CHAPTER 13)

1 Explain the important differences between the thermoplastic and thermosetting types of plastics materials. Give one example of each type.
2 Give an account of the vinyl plastics and their use in building.
3 List the advantages and disadvantages of polythene as a material for use in water supply and distributing installations.
4 Make a list of the important plastics materials used in building and summarise their principal properties and uses.

EXERCISES (CHAPTER 14)

1 Explain the following terms used in connection with adhesives: (a) pot life (b) shelf life (c) gap-filling (d) thermosetting (e) catalyst.
2 Discuss the principal advantages and disadvantages of synthetic resin adhesives in comparison with animal glues.
3 Distinguish two types of building mastics in relation to their physical properties.
What are the principal physical and chemical properties required of a building mastic?

EXERCISES (CHAPTER 15)

1 What are the main functions of paints and paint treatments?
2 By what action do the following types of paint film dry or harden (a) oil paints (b) emulsion paints (c) solvent-type paints (d) water paints (e) synthetic resin paints?
3 State concisely the function of a primer in a complete paint treatment.
Discuss the merits of the various types of priming paints used on metals, giving examples of their suitability for application to particular ferrous or non-ferrous metals.

4 Get information and write notes on the following protective treatments for metals: (a) Bower Barffing (b) Sheradising (c) Phosphating (d) Metallisation.

EXERCISES (CHAPTER 16)

1 Explain the meaning of the following terms in their application to building: (a) asphaltic cement (b) natural rock asphalt (c) mastic asphalt.
Discuss the various uses of mastic asphalt in building, mentioning any advantages or disadvantages which it has in relation to alternative materials.

2 Name the materials commonly used in the manufacture of roofing felts.
Distinguish between 'saturated' and 'coated' felts.
Describe briefly the following types of roofing felt and indicate their normal uses: (a) reinforced (b) mineral-surfaced.

3 What raw materials are used in magnesium oxychloride flooring? What precautions are required in preparing, mixing and applying the material?

EXERCISES (CHAPTER 17)

1 List the raw materials used in the manufacture of glass. Name and describe one method used in the manufacture of flat sheet glass.

2 Explain the meaning of the following terms when applied to glass for glazing (a) transparent (b) translucent (c) opal.
In what particular respect is polished plate glass superior to drawn sheet glass?

3 Write notes explaining the following: (a) float glass (b) wired glass (c) toughened glass (d) lead (X-ray) glass.

EXERCISES (CHAPTER 18)

1 Compare and contrast the properties of cast iron and steel, with reference to their different composition.

2 What particular properties of aluminium and its alloys makes this group of metals useful in building?

3 Various grades of steel tube are supplied for installations conveying gas, water or steam. Obtain and give the following information: (a) what are these grades, and how are pipes marked to make the grades distinct? (b) which of the grades is used principally for water, and which for gas? (c) what finishes may be specified when ordering steel tube?

4 Find out, and state, in what forms copper tube is supplied for water, gas and sanitation, and in what gauges.

TEST PAPERS WITHOUT ANSWERS

1 Name three types of Portland cement, explain the difference in
their properties and state when each would be used.

2 Explain the setting action of gypsum building plasters.
Why are hemihydrate plasters generally less hard and dense than
anhydrous types?

3 Explain concisely the difference between a Portland cement, a
lime and a plaster, as used in building.

4 Name any four lightweight aggregates used in building work.
What advantages are obtained by the use of lightweight aggregates in mortar, plaster or concrete?

5 What general principles should govern the choice of mix proportions for a mortar?

6 Explain the purpose of making a slump test on concrete.
Describe the slump test and give a sketch of the apparatus used.
Name and illustrate with a sketch the three types of slump which
can occur.

7 A concrete mix is specified to contain the following amounts of
dry materials and water: cement 180 kg, sand 300 kg, gravel 540
kg, water 90 litres. The total moisture content of the sand used is
5 % by dry weight and that of the gravel 2% by dry weight. (a)
Calculate the amounts of damp sand and gravel to be weighed
into the mixer (b) State how much water should be added at the
mixer.

8 Describe the current BS method of testing clay building bricks
(with and without frogs) for crushing strength.
What minimum crushing strength would you expect from a
clay engineering brick?
A specimen measuring $8\frac{1}{2} \times 4$ in (215×102 mm), when tested in

compression, failed at a load of 45 tonf (45·7 tonnef, or 448 kN)
Calculate the failing stress, in lbf/in², in kgf/cm², and in N/mm².

9 Explain why it is generally much easier to cut and dress a
sedimentary stone than a granite.
What is 'quarry sap', and how does it affect the cutting and
dressing of masonry?
Suggest what properties are important in a natural stone which
is to be used as a crushed rock aggregate in precast paving flags
and kerbs.

10 Discuss general measures which may be taken to prevent the
decay of timber.

11 Give an account of the uses of plastics materials in building.

12 What is the primary function of each of the following in a
standard paint treatment: (a) primer (b) undercoat (c) finish.
Specify a suitable paint treatment to be used externally for each
of the following where a coloured finish is required: (a) a softwood ledged and braced gate (b) ferrous metal fencing (c) concrete posts (d) a bitumen-coated soil pipe (e) galvanised
corrugated steel sheet.

13 Discuss the principal differences between pitch fibre and stoneware pipes for drainage and surface water.

14 For what main purposes and in what forms is copper used in
building?
What particular type of copper sheet would you specify for
roofing work?
Why does copper not require painting when used externally?
Copper is said to be ductile but not subject to creep—what
does this mean?
Explain why zinc sheet should not normally be used in contact
with copper.

TEST PAPER 2

1 Explain by what process a Portland cement sets and hardens.
Account for the difference in setting and hardening times of
ordinary Portland, rapid-hardening Portland and extra rapid-
hardening Portland cements.
Why is gypsum added to Portland cement during its manufacture?

2 Name the British Standard laboratory tests which may be applied to gypsum building plasters.

Outline the method of any two of these tests, and indicate how the results are established and assessed.

3 Explain the difference between (a) chalk and limestone (b) quicklime and slaked lime (c) hydraulic lime and non-hydraulic lime.

4 Name any three standard field tests applied to aggregates, and three different tests which may be applied in a laboratory.

State the purpose of each test which you mention.

Describe in detail one field test for aggregates and one laboratory test.

5 A coarse and a fine aggregate have the gradings shown below. Determine the combined grading when using 22% sand.

Also find the percentages of each for a combined grading with an amount of 35% passing the $\frac{3}{16}$ in (4·76 mm) mesh sieve, and state the overall grading which results.

Sieve size	$\frac{3}{4}$ in	$\frac{3}{8}$ in	$\frac{3}{16}$ in	No. 7
% passing:	(19·05 mm)	(9·53 mm)	(4·76 mm)	(2·40 mm)
Fine			100	92
Coarse	100	31	7	0

Sieve size	No. 14	No. 25	No. 52	No. 100
% passing:	(1·20 mm)	(600 μm)	(300 μm)	(150 μm)
Fine	76	48	20	3

6 What factors govern the durability of a brickwork mortar, and in what way?

7 A concrete mix has an aggregate:cement ratio of 4·8 by weight with a sand content of 30%, and a water:cement ratio of 0.50. The specific gravities are; cement 3·10, fine aggregate 2·60 and coarse aggregate 2·65.

Calculate the amount of cement required in kilograms per cubic metre of the fully compacted finished concrete.

8 The two governing factors in the design of a dense concrete mix for structural work are the water:cement ratio and the aggregate: cement ratio. State which of these is the critical factor in relation to the following properties of the concrete, and indicate the

effect on that property of an increase in the ratio concerned: (a) workability (b) crushing strength (c) durability.

Use standard data to design a concrete mix to meet the following requirements: minimum strength 4000 lbf/in² (281 kgf/cm² or 27·5 N/mm²) at 28 days, using rapid-hardening Portland cement where site production control is very good and includes weight batching. Aggregates are $\frac{3}{4}$–$\frac{3}{16}$ in (19–4·75 mm) graded crushed rock to BS 882 from an efficient quarry, and BS Zone 3 sand from an efficient pit. Medium workability is required, and the degree of exposure for the construction is normal (temperate climate).

9 Name two materials other than clay used in the manufacture of roofing tiles.

Name two BS tests applied to clay roofing tiles, and give an outline of the methods of these tests.

10 Explain how the composition and characteristic physical properties of *either* sandstone *or* limestone influences its selection and suitability for use in building work, making particular reference to durability.

11 Name two insects which commonly attack timber during their life-cycle and state (a) the effects which they produce on infested timber (b) the characteristics of their species which enable them to be identified.

12 Compare and contrast the properties and uses of plastics materials and metals in building.

13 Write notes to explain the following terms used in paint technology (a) alkali-attack (b) blistering (c) chalking (d) bleeding (e) sulphiding.

14 Name the main uses of lead sheet in building.

Describe the main advantages and disadvantages of lead sheet as a building material.

ANSWERS TO CHAPTER EXERCISES

Chapter 4

4 5·32, 31%

Chapter 11

3 25%, 13·2 kgf

Chapter 8

4 65 l

5 Cement: fine: coarse/water
   (a) 180 kg: 423 kg: 766 kg/108 kg (or l)
   (b) 180 kg: 444 kg: 781 kg/72 kg (or l)

INDEX

ACCELERATORS, in concrete, 137; in mortar, 87
Adhesives, 238–44; animal glues, 239; bitumen, 242; casein, 239; mastic, 239; paste types, 241; PVA types, 241; silicate, 242; solvent types, 233, 242; staining by, 242; synthetic resins, 234, 240; tests on, 242–4
Aerated concrete, 139
Aggregates, 51–81; artificial, 52; blast-furnace slag, 52, 291; bulking of, 66, 73, 118; clinker, 52; combined grading, 60; continuous grading, 114; crushed stone sand, 52, 84; diatomaceous earths, 53; exfoliated vermiculite, 53; expanded clay, shale, slate, 53; expanded perlite, 53; fine, 52, 114, 115; fineness modulus, 58; foamed slag, 52; for concrete, 52, 113; for floor screeds, 52, 99; for mortar, 52, 86; for rendering, 52, 98; gap-graded, 138; grading, 54, 98; impurities in, 68, 74; lightweight, 52, 99; moisture content, 68; pulverised fuel ash, 53; pumice, 53; quality of, 68; sampling, 54; sawdust, 53; silt, 68, 74; soluble salts in, 68; surface area index, 60; tests on, 69; types, 51–4; water absorption, 73; wood fibre, 53

Air-entrained concrete, 140, 143
Alkyd resins, 234, 252
Alloys, eutectic, 290; of aluminium, 295; of copper, 296; of lead, 297; of steel, 293
Aluminium, 294; paints, 256
Aminoplastics, 232
Anhydrite, 31
Anhydrous gypsum plaster, 30–1, 105
Anodising, 296
Asbestos-cement sheets, 225; insulating board, 225; partition board, 226
Asphalt, 269; mastic, 270; rock, 269; tests on, 271
Asphaltic bitumen, 269

BARYTES, as aggregate, 53
Bauxite, 28, 148
Bitumen, 269; adhesives, 242; emulsions, 269, 273–4; sealing, 256; tests on, 271
Bituminous asphalt, 269–70; materials, 269–76; paints, 251; tests on, 271
Blast-furnace slag, 52, 291
Blockboard, 218, 222
Blocks, clay, 153; concrete, 161; glass, 280; lightweight, 161; tests on clay, 154
Bonding agents (see also *Adhesives*), 94, 98, 100, 106
Brass, 296

349

surface moulds, 202; tests on,
209–13; tracheids, 186, 197;
tyloses in, 194; vessels, 192;
wane in, 207; wet rot, 201; white
rots, 202
Tree, classification, 188; growth,
182
Tyrolean, 96

UNSOUNDNESS, 21

VARNISH, oil, 250, 254; resin,
252; spirit, 254
Vebe consistometer test, 135
Vermiculite, in plasters, 30, 33
Vicat apparatus, 19
Vinyl plastics, 232
Vitreous china, 169

Voids, air, measuring, 140; in
concrete, 109, 115

WALL tiles, ceramic, 167
Wallboard, asbestos, 225; fibre
building, 223
Ware, sanitary, 168–9
Water-glass, 277
Water, paints, 250; repellents, 174
Wet dash, 96; rot, 201
Window putty, 245
Wired glass, 281
Wood—see *Timber*
Wood wool slabs, 223
Workability, of concrete, 109, 112,
130; tests, 130–5; of lime, 47;
of rendering mixes, 93
Wrought iron, 292

ZINC, 298

SCIENCE IN BUILDING
for Craft Student and Technician
3
MATERIALS